AF531312

# FUNDAMENTALS OF ANIMAL BEHAVIOUR

# FUNDAMENTALS OF ANIMAL BEHAVIOUR

*By*
**Dr. Amita Sarkar**
*Lecturer*
*Department of Zoology*
*Agra College*
*Agra (U.P.)*
*(India)*

DISCOVERY PUBLISHING HOUSE
NEW DELHI-110002

First Published-2003

Reprint–2004

ISBN 81-7141-742-6



*Published by*

**DISCOVERY PUBLISHING HOUSE**
4831/24, Ansari Road, Prahlad Street,
Darya Ganj, New Delhi-110002 (India)
Phone: 23279245 • Fax: 91-11-23253475
E-mail:dphtemp@indiatimes.com

*Printed at:*

Tarun Offset Printers,

Delhi–110 053

# Preface

The present title Fundamentals of Animal Behaviour aims to stimulate an understanding of Behaviour of wide variety of animals including man, farm animals and best species to inspire the reader to take an interest in the field. The text integrates the descriptive and experimental approaches into a conceptual framework for the analysis of behavioural studies. It is profusely and attractively illustrated with line diagrams.

It introduces the basic ideas and concepts of modern ethology set in historical context, thereby showing how views have changed since the simple theories put forward by the founders of the field, such as Lorenz and Tinberg 40 years or more ago.

The title is not intended to be comprehensive, nor could it be at this length, but it concentrates as putting across the basic principles of the subject as briefly and lucidly as possible. It does this with the aid of carefully selected examples some recent and others classic or the field, and with numerous illustrations. The aim is to enthuse the reader with the active and exciting area of research and to lay a solid foundation on which further study of its various facts may be based.

The author has freely consulted various articles, discussion notes, reviews and extracts from the various scientific journals in the preparation of the present book, in able to make it comprehensive and upto date.

Though every care has been taken by the printer, publisher and me, it is quite likely that some errors might have found their way into the book but I hope these are of very insignificant nature. However, suggestions to improve book and pointing out of errors and mistakes, if any, will gratefully be appreciated.

The author expresses grateful to her friends and colleagues whose constant inspiration have initiated her in bringing out this book.

Special thanks are given to Mr. Wasan and staff of M/s Discovery Publishing House for their whole hearted co-operation in the publication of this book.

**Author**

# CONTENTS

# 1

# PRINCIPLES OF BEHAVIOUR

The behaviour of animals is flexible and diverse. During a typical day an individual animal may perform, in a particular order, dozens of different behaviour patterns, or units. A male bird in the spring, for example, may sing at dawn, and then feed (using many different foraging techniques), hide from predators, defend its territory, and court a female, during one day. Ethologists have not progressed far in describing the mechanisms that control the complete behavioural repertory of an animal through its life–that is, the way it decides what behaviour patterns to perform in a given set of circumstances, how to trade off its motives to perform one activity rather than another, and by what physiological mechanisms these decisions are made. If ethologists have not progressed far in solving the total problem, they have made many important and curious discoveries about the mechanisms controlling some individual behaviour patterns. This is particularly true of animal senses. The general, elemental mechanisms controlling behaviour are known, even if how they are combined to produce the full behavioural output of an animal is not. Muscular contractions are an important mechanical source of many behaviour patterns. Animals control their muscles by their nervous systems.

The nervous system also carries information from the sense organs and, by mechanisms of excitation and inhibition among its different parts, integrates the animals's senses with its internal tendencies and preferences. But the study of animal nervous systems has not in all cases turned out to be the easiest route to understanding how they control their behaviour, and we shall consider some examples of how animal senses have been successfully studied at a higher level than that of sensory neurons, and their behavioural choices at a higher level than the influence of neurons on one another.

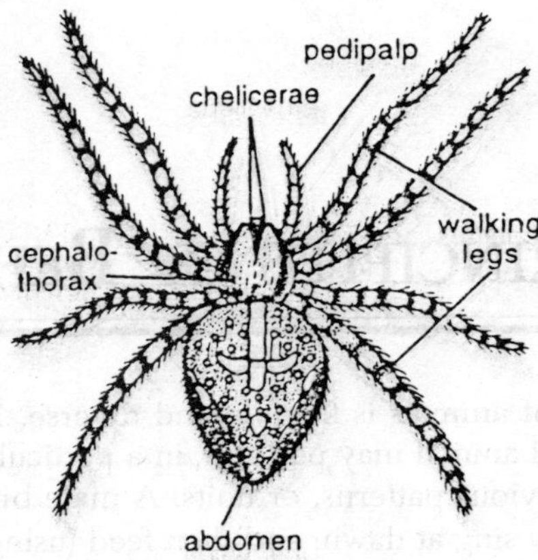


*Fig. 1.1. Aranea (Spider).*

## The Spider's Web

Most spiders are solitary, with one individual living an each web a wondering as solitary hunters. Among spiders that build sheet like webs–the kinds so after seen on lawns and in hedges–Here are species in which the spider lings remain as the web and are fed by their another. In some species the spider lings build their own small webs next to their mothers. The final stage in the evolution of sociality is communal construction of the web. Several spiders may run out to subdue a large prey. The nature of these social systems strongly suggests that they evolved via the familial route. Among orb-weaving spiders, however, sociality appears to have volved via the parasocial route. Individual build their own webs in most species.

The start of sociality was probably clumping of webs in unusually favourable sites. This led in a few cases to the joining of webs and construction of communal retreat. Each spider builds and occupies her own web and there is no communal subduing of prey. There is no care of the young, overlaping of generations, or division of labor. In the social orb weavers there is no evidence that the females are closely related to one another. Perhaps the araneid spiders, such as the common garden spider *Araneus*

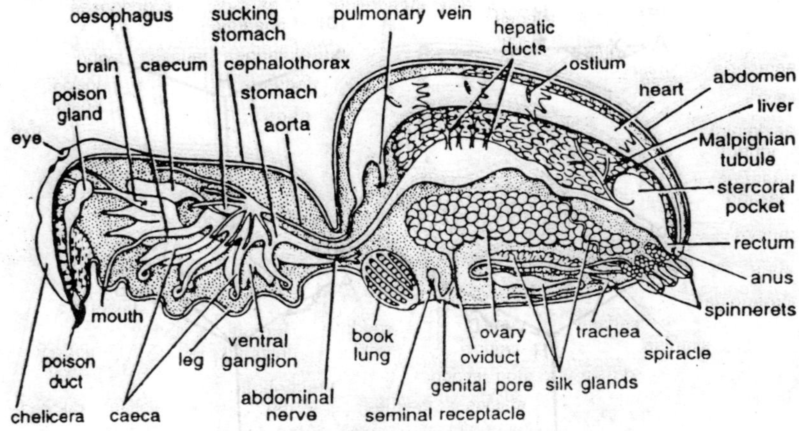


*Fig. 1.2. V.S. of spider showing anatomy.*

*diadematus,* which do build orb webs, have a concept of the web; we cannot tell. However, observations of spiders in action suggest that they follow a series of rules, which in themselves would be sufficient to lead the spider to build a web even if it had no idea of what the end point should be.

Not all the flexible details of the spider's building program have been worked out, but the main rules were discovered simply by watching them build. The observations are easy to make. Orb webs are spun by common spiders all over the world; and the whole operation takes less than half an hour. The end product, the orb web, is a regular structure made up of frame, radial spokes, and catching spiral. The stages of its construction were first formalized by *Hans Peters* in 1939, and they are as follows. The spider starts with the frame and spokes. To begin with, she spins a thread with one unattached end and allows it to be blown by the breeze. The other end is attached to the spider, by the spinnerets of the abdomen, which are the source of the thread. The loose end will soon become entangled in some object, such as a nearby twig. The spider then bites through her end of the thread and, having attached a new thread at her point of departure, walks of down the wind-cast thread, spinning another thread as she goes. When she arrives at the other end, she attaches the new thread. She then turns, walks some distance up the thread and attaches a new thread there. She now allows herself to fall, spinning a thread behind her, to a third attachment point (C). The Y-shaped structure provides

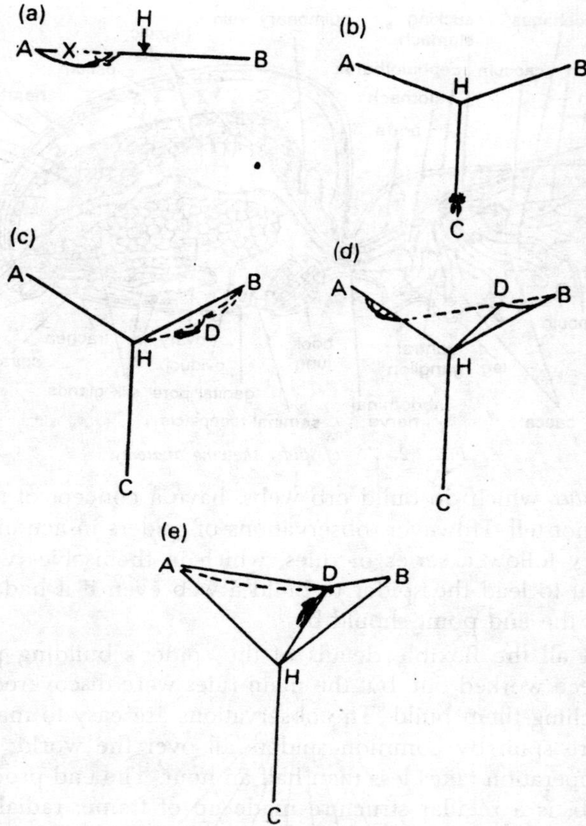


*Fig. 1.3. The stage in the construction of an orb web.*

the scaffolding for the web. She next builds by turns the radial spokes and frame. She returns to the centre H (which stands for hubs, as the centre of the completed web is called), attaches a thread, walks to the outside (B) spinning a thread as she goes, and attaches the other end of if at B; she then walks back down the thread to a, point D where she attaches a new thread. Now, after she has walked on to A, she has constructed a frame from B to A and a radial spoke from D to H. The same sequence of movements final web the angles of the radii at the hub are rather constant, at about 15.'

The radii are built in a regular order, the next spoke always being added in the largest unfilled sector of the web. In other words, the spider measures the angles between the existing spokes at the hub, and chooses to spin the next spoke between the two spokes subtending the largest angle. One she has returned to the hub after building a spoke she pulls each spoke with her front legs, which may be how she measures the angles, perhaps by estimating the tension in each spoke. At all events, sight is unnecessary for correct building because spiders can build normal webs in complete darkness, or when sightless. Following the rule for where to spin the next spoke results in spokes being laid down in the kind of order illustrated in Figure, with each new spoke in a different sector from the previous one. Once the frame and radii are complete (The formation of frame and radii takes about five minutes), the spider starts on the spiral. She builds it in two stages.

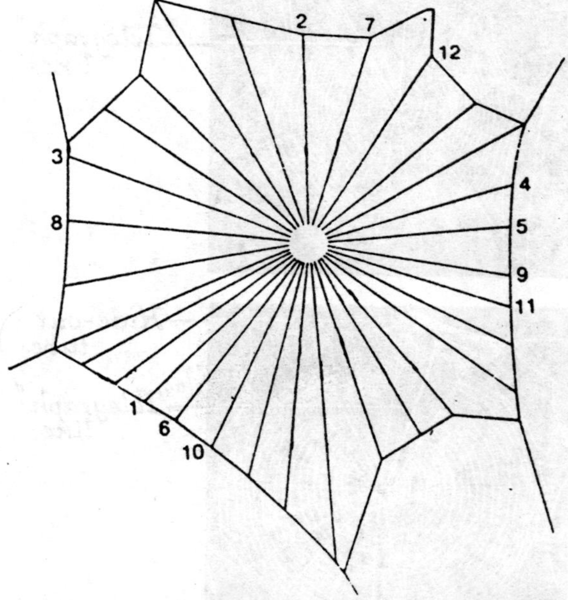


*Fig. 1.4. An orb web, with most of its radial spokes, but no catching spiral. The numbers indicate the order in which the radii are laid down.*

Starting from the hub, she first spirals outwards, to lay down an *'auxiliary spiral.'* She does this by walking outwards, is an arithmetic spiral, spinning and attaching a thread behind her, until she reaches the edge. The auxiliary spiral is a temporary structure.

For at the outside she turns round and, working inwards now, spins the sticky she cuts the auxiliary spiral, the function of which was to act as a guideline for laying down the catching spiral. The whole web is finished once the sticky catching thread has spiralled into the hub. Actually, the spiral is not a perfect spiral; it is asymmetrical. Orb webs are not typically circular. They are elongated in the bottom half, with the hub above the centre. The bottom is lengthened by means of a ladder of sticky switch-backs in the bottom sector of the web, which can be seen in Figure. Only after a few turns of the ladder does the spider begin the true spiral. The main rules of the orb web spider are first to build a Y-

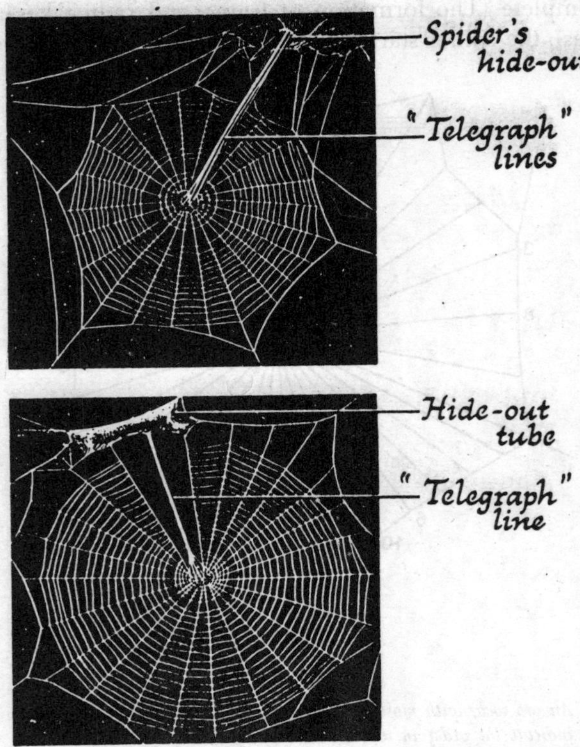


*Fig. 1.5. Four types of orb web characteristic of different species of spiders. Top to bottom: in Araneus, the spider's "hideout" is connected to the orb by "telegraph" lines. In Zygiella, the lair is in the framework of the orb and connected to the center by a strong signal line between two radii without spiral threads. In Argiope, the black and yellow garden spider, the white stabilimentum is directly above and below the orb center. Metepeira weaves and irregular hub and a labyrinthine snare above the orb.*

shaped scaffold, then, in a set order, the frame and radial spokes; and finally the auxiliary and sticky spirals.

Each main rule contains a set of subrules for measuring angles and walking set distances up certain threads. By following the program of elementary rules the spider can-build a complex structure without having a plan of it in her head. This is not to say that the spider lacks such a plan. Spiders surely are more complex machines than this account has admitted. There is probably more to the control of web building than the series of rules considered here, But this fact does not principle which the spider's web illustrates: that apparently complex structures can be build up from a fairly simple set of rules. The principles is of general importance. It is often possible to break down the apparent complexity of animal behaviour by analysing it into a series of 'rules of thumb' that the animal itself could, in some form, be following. Searching for the rules is a powerful method of studying the control of behaviour, because it is easier to pursue research on simple rules than one complex ones.

It is easier to follow up the nervous control of a leg movement than of conceptual thought. Of course, complex systems can be studied too; but because it is more difficult it simply takes longer. More sophisticated rules are most easily studies if they are added piecemeal on to a concrete foundation of knowledge. We therefore, advance fastest in our understanding of behaviour if we search for the simplest hypothetical mechanisms. Ethologists do not invoke less simple mechanisms–of conceptual thought or conscious calculation, for example–if they are not needed to explain the behaviour that is being observed. So far we have thought of an animal as executing a series of rules. But this execution must in its turn by controlled by some mechanisms. The controlling mechanism ultimately is the nervous system; and to see how that works we must change our study animal, from the araneid spider, to squids and crickets.

## The Nervous System

It is the integrated system and is composed of a network of nerve fibres. A nerve fibre is are bundle of cells called *neurons.* A muscle contracts when the neuron attached to it, called a motor neuron, becomes electrically active. Neurophysiologists first worked out how neurons work in one particular kind of neuron which is found in the squid. When a squid senses that an enemy is nearby, it suddenly contrasts its body, forcing water out of the area called its *mantle cavity*, and jets away. The squid may also discharge a screen of ink to cover its escape.

The 'escape reaction', as it is called, is controlled by a large (and therefore easily studied) neuron, which extends in stages from just outside the squid's brain at one end of its body, to the muscles, the contraction of which at the other end of the body effects the escape reaction. The impulse travel in the nerve fibre in the electrochemical means. The neurosome is polarized at rest and has positively not as well as negatively charged ions on different sides of the membrane. When the neuron is activated, positively charged sodium ions rush into the neuron. The inside of the neuron in the activated area becomes electrically positive relative to the outside.

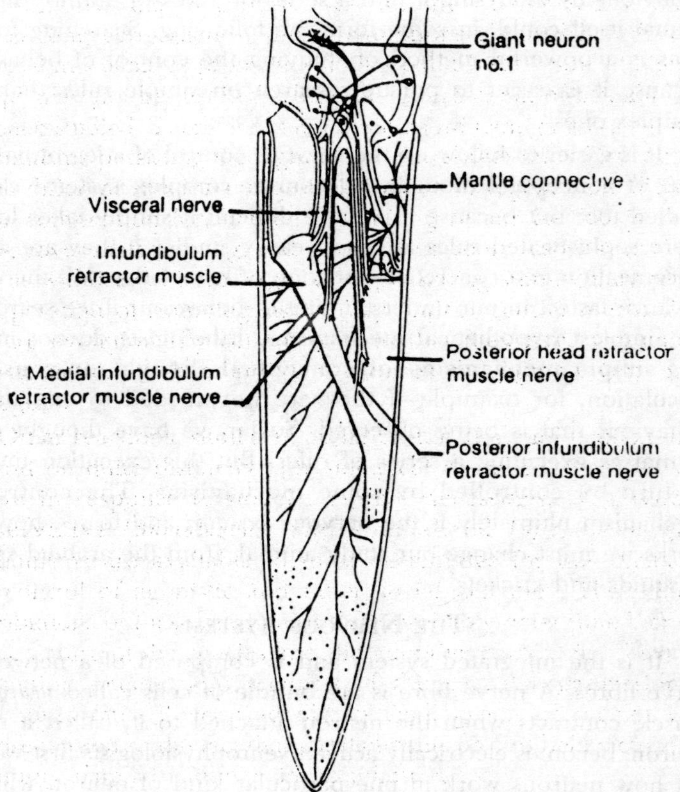


*Fig. 1.6. The course of the giant neurons of the squid.*

This pulse of depolarization, as it is called, travels rapidly along the neuron until it reaches the muscle. The electric current then stimulates the muscle to contract. But what stimulated the motor neuron to begin with? We must turn to the other end of the neuron, away from the muscle, where the motor neuron is joined to many other neurons at a junction called a *synapse.* Here is the origin of motor neuron activity.

Chemical neurotransmitters are released by the neuron at the synapse, and if sufficient is released, the motor neuron will fire into action. The motor neuron is thus controlled by the activity of the many neurons preceding it in the nervous system, whose activity is in turn controlled by the central nervous system in the brain. Sensory input from the external environment is also carried to the brain encoded in electrical depolarizations of neurons, called *sensory neurons,* which pass from sensory organ to the brain. Speaking very generally, the behaviour of an animal is mainly directed by its brain, which integrates the actions of thousands of neurons, but by processes that are too little understood to be worth dwelling upon. The neuronal control of some simple behaviour patterns is, however, quite well understood. The singing of crickets is an example, and it will serve to illustrate how the nervous system produces behaviour. Singing in crickets is not a response to an external stimulus. It is influenced by external factors such as temperature; but is not a direct response to them. It is caused by the endogenous action of the nervous system. Male crickets sing, by scraping their wings together, to attract females.

The movement of wings is controlled by a set of muscles in the thorax (the thorax is the middle section of the insect, between the head and the abdomen). Motor neurons run from the thoracic ganglion (a ganglion is a mini-brain, an aggregation of many nerves) to the singing muscles. It takes many muscle fibres to move a wing, and each muscle fibre has its own motor neuron. All the muscle fibres must contract at the same time, and the co-ordination is produced by a 'command interneuron', onto which all the motor neurons join. Soon after the command interneuron fires, all the motor neurons fire in unison.

The co-ordinatory function of the command interneuron is situated fires, all the motor neurons fire in unison. The co-ordinatory function of the command neuron has been proved by experiment. The command interneuron fires, all the motor neurons fire in unison.

The co-ordinatory function of the command neuron has been proved by experiment. The command interneuron is situated in the cricket's neck. A neurophysiologist can locate it and join it to an electricity supply. If the electricity is switched on, all the motor neurons fire together, and the cricket produces a perfect song. It will produce its song even if it has had its head cut off. The command neuron, motor neurons, muscles, and wings form a complete singing machine.

## Sense Organs or Receptors

All animals live a fluctuating environment and they need to be able to find their way around their environment, find food, recognize the species and sex of other individuals, even (in some cases) whether another individual is a member of the same, or a different, group, and detect the signals sent to them. For all these functions they rely on their sense organs. Different kinds of animals have different sets of sense organs. The set of sense organs possessed by each kind of animal is appropriate to the environment in which it lives. For example, species of fish and shrimp which live in dark, underground caves do not have eyes, or have eyes so reduced that they no longer work; there is no advantage in having light-sensitive organs where there is no light. The human set of eyes, ears, touch, and relatively poor taste and smell is just one particular, not very common, set of sense organs; most species of animals live in a world of senses very different from our. The various receptors of animals can be divided into three groups: *exteroceptors, enteroceptors* and *proprioceptors.* This division was suggested by the neurophysiologist *Charles Sherrington* at the beginning of the century. Exteroceptors sense the state of the environment outside the animal; enteroceptors the state of the animal's body; proprioceptors the animal's movement by sensing the position of its muscles.

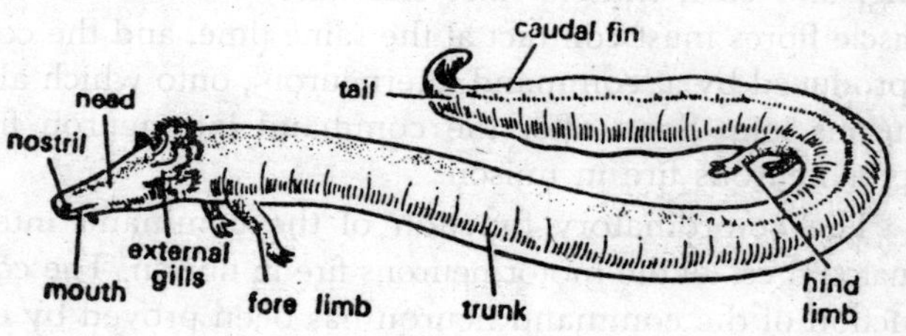


*Fig. 1.7. Proteus.*

In mammals, enteroceptors include organs that sense the body temperature and chemoceptors that sense the concentrations of chemicals such as hormones and carbon dioxide. The classic five human senses–sight, hearing, touch, taste, smell–are effected by exteroceptors. In fact there is no perfect classification of these external senses. Taste and smell are both chemical senses (as are many of the enteroceptors). Hearing and touch are both mechanical senses. Other species possess other senses which do not easily fit into the five-way division. But, bearing in mind that the classification is crude, we can consider quickly some examples of each kind.

## Galvano Receptors

The electrical sense is lacking in humans but studied in fishes where it is greatly developed. Their electrical sense is not the same as the sensitivity to pain by which we become aware of an electric shock, but another sense, comparable to our sense of hearing or smell. Some kinds of fish, such as dogfish (which is a member of the elasmobranch group that includes sharks and rays), use their electric sense to find food buried in the bottom sand by the disturbance to the electric field that the buried living matter causes. The electric sense organs of elasmobranch fish are called the *ampullae of Lorenzini* and are a set of jelly-filled tubes beneath the skin. Other animals, such as a honey-bees, and bacteria, can sense magnetism. Experiments we shall meet later suggest that at least some kinds of birds can also sense the magnetic field.

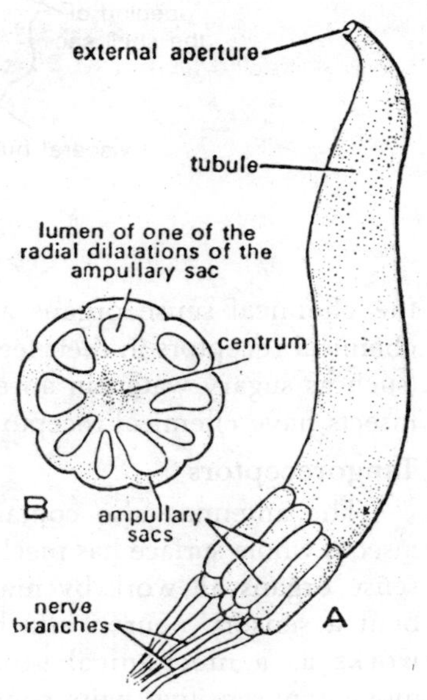


*Fig. 1.8. Scoliodon. A–An ampulla of Lorenzini. B–Section of an ampullary sac.*

## Chemoreceptors

Chemical sense organs are found in many place in different kinds of animals. The sea hare, which is a favourite animal for work on the nervous system, can smell seaweed, on which it lives. By recording the activity of the nervous system at different parts of the animal while it is smelling seaweed, it has been found that

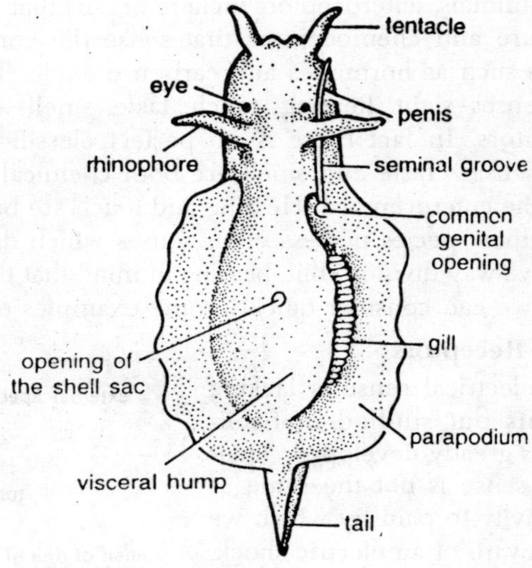


*Fig. 1.9. Aplysia.*

the chemical sense organs are in its tentacles. Houseflies have chemical receptors in their feet, which enables them to detect food (such as sugary water, in an experiment) by walking into it. Most insects have chemical receptors in their antennae.

**Tangoreceptors**

The antennae also contain mechanical sense organs; but an insect's whole surface has mechanical sense organs on it. Mechanical sense organs all work by means of tiny hairs. When the hair is bent a sensory neuron attached to it fires into action. Hearing works as a mechanical sense, and is also effected by small movement-sensitive hairs connected in some way (depending on the species) to a membrane that is set in oscillation by sound. Fish and some amphibians possess special organs for detecting water pressure called the *lateral line organ.* The lateral line is a channel under the skin of each side of the animal, with little holes leading to the outside. The flow of water into the lateral line allows the fish to measure to movement of water with respect to itself.

**Photoreceptors**

The final class of sense organs is the light receptors. Eyes are found is more or less complex form in many kinds of animals. But eyes are not the only light-sensitive organ known in nature.

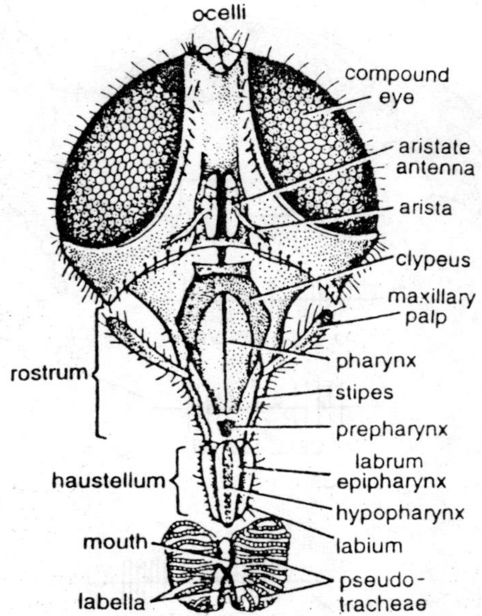


*Fig. 1.10. Housefly. Head and mouth parts (Dorsal view).*

Insects, for instance, have three light sensitive ocelli on the top of their heads, behind their compound eyes. The functions of the ocelli are uncertain.

## Echolocation in Bats

Bats are the aerial mammals showing volant adaptations. There are over 900 species of bats. The group has perhaps proliferated because its echolocatory sense allows it to live in an environment containing almost no competitors. They fly by night, feeding on night-flying insects, particularly moths. Bats can fly with ease in complete darkness; they do not collide with obstructions, and they catch their prey on the wing, They are capable of flying around a laboratory room which is criss-crossed with a network of wires of a diameter of as little as 3 hundredths of an inch, and catching flying moths from a range of 8 feet. How do bats achieve this? The greatest experimental biologist of the eighteenth century, the Italian *Lazzaro Spallanzani,* could not solve the problem. He did find that if he stuffed up the bats ears their ability to avoid obstructions declined. But he did not know how to explain this result. Bats remained a puzzle until after advanced equipment of

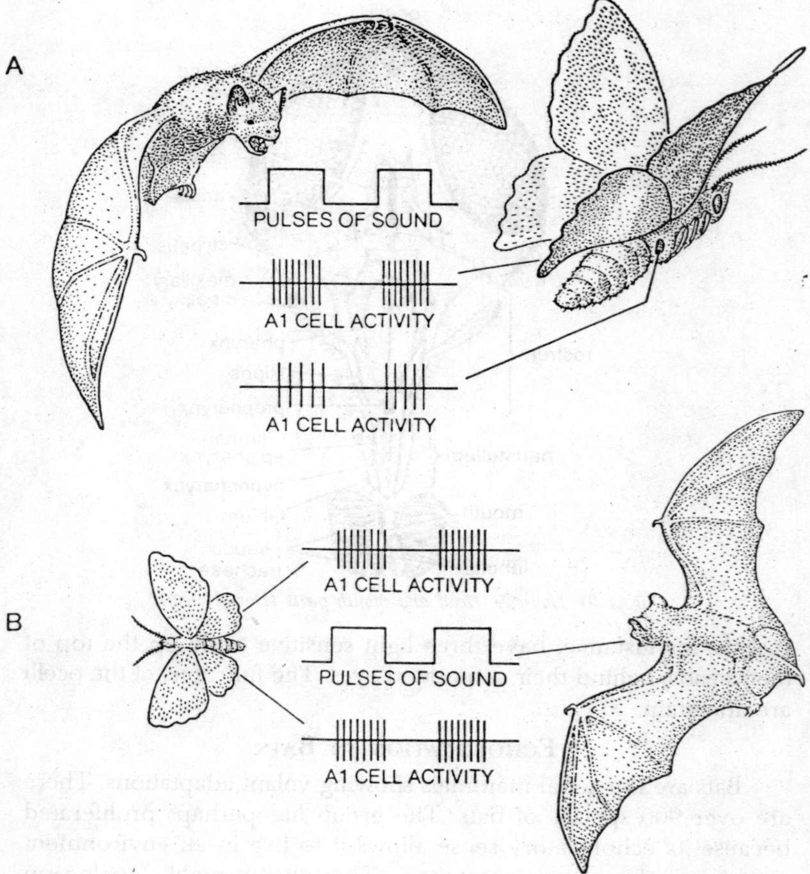


*Fig. 1.11. Activity in the A1 receptors of a moth's ears on detecting the cries of bats located in different points in space. A–A bat is to one side of the moth; the receptor on the side closest to the predator fires sooner and more often than the shielded receptor. B–A bat is directly behind the moth; both A1 fibers fire at the same rate and time.*

recording and producing sound has been developed along with radar in the Second World War. The equipment was applied to the bat puzzle by *Donald Griffin* in the 1950s. He proved that the little brown bat finds its way around by listening to the echos of high-pitched sounds that it makes itself. Dogs can hear sounds of the higher frequency of about 40,000 Herfa. Humans cannot hear sounds much outside the frequency range 2000-20,000 Hertz (Hz).

(The higher the frequency the higher the pitch). Bats make sounds mainly of 20,000 Hz and higher–some emit sounds of up to 100,000 Hz–and most bat sounds are therefore inaudible to humans. One advantage to the bat of using such high-pitched sounds is that it is undisturbed by background noise (most noises in nature have a frequency lower than 20,000 Hz). By concentrating on high-frequency sounds, bats live in a world silent except for their own noises.

It is essential to the bat that there should be no interfering background noise. In an experiment, *Griffin* blasted noises of frequencies of more than 20,000 Hz into a room with obstructions. Flying bats now bumped into the obstacles and fell into the floor. As well as showing the importance of a silent background, this experiment also gives part of the evidence that bats use high frequency sounds to find their way around. Echolocation works by measuring the time interval between releasing a short pulse of sound and hearing its echo. The longer it takes the echo to come back, the further away the object must be. The sound pulses have to be very short, to prevent overlap between the emitted sound and its echo: that bat can hear itself as well as the echo.

The pulses of sound are indeed very short, the little brown bat releases four or five distinct pulses every second during ordinary flight. It increases the rate of pulses when it detects on object, and as it flies closer to the object its pulse rate rises further. It does so because, as it approaches the object, the delay until the echo comes back becomes shorter and shorter. It therefore, makes it pulses shorter, again to prevent overlap between pulse and echo. Echolocation becomes ineffective beyond distances of about 30-40 metres, because sound is rapidly absorbed in air. But even for these short distances the bat must listen for a very faint echo of its much louder original pulse; the sound pulse emitted by the bat is 2000 times louder than the echo. The bat therefore has the problem that it has to make a loud noise and then hear a soft noise immediately afterwards. When an ear has been blasted with a loud sound it becomes less sensitive to recover after listening to a very loud noise. Bats have a number of methods of solving this problem. One is to make the ear less sensitive when emitting the sound.

A nerve going to the muscle of the ear is automatically activated whenever the bat releases its sound pulse. This nerve causes the bat's ear to relax about 5 thousands this of the a second before the pulse is emitted, and to recover about 10 thousandths of a second later. The ear will then be a peak sensitivity for picking up the echo. Bats are not the only kind of animal to use echolocation. They are however, by far the most extensively studied. Other animals which echolocate are the dolphin and other 'toothed' whales, small mammals called shrews, and a bird which lives in dark caves called the Malayan cave swiflet.

## Use of Sensory Informations

Animal behaviour is not only a set of simple responses to various environment situations; but we can make some points about how an animal uses its sensory information without implying that it is all there is to behaviour. All sense organs send their knowledge in coded form to the central nervous system by means of sensory neuron. Different sensory neurons are stimulated by different properties of the environment. Light sensitive neurons, for instance, contain a light sensitive pigment, which, when illuminated, changes its chemical form and causes the neuron to burst into action. A special neuron is required to sense light: if you shone light on any other kind of neuron it would have no effect on it; only a neuron containing a light sensitive pigment is set in action by light. Most behaviour patterns are controlled in the central nervous system which integrates the information from the sense with other neuron systems, to control the animal's behavioural output.

Some behaviour patterns are not centrally controlled. They are called peripheral reflexes (of which the human knee jerk is an example), and are controlled by a simple system of one sensory neuron and one motor neuron. The sensory neuron connects directly on the motor neuron. The sensory neuron connects directly on to the motor neuron, which in turn controls the muscles that effect the behaviour pattern. When the sensory neuron fires (after mechanical stimulation in the case of knee jerk reflex), it stimulates the motor neuron, which causes certain muscles to contrast.

We shall meet some other examples of reflexes; but it most behaviour patterns the sensory input and behavioural output of the animal are less directly connected, and the senses exert their influence on behaviour through the central nervous system. There is potentially an enormous amount of information about the environment, far more than an animal could make use of in deciding on a course of action. The environment, for instance, is brim-full of electromagnetic rays: it has a constantly changing pattern of light and dark, different colours, and a continual hum of x-rays and radio waves, together with high and low frequency radiation of which we are not normally aware.

Most of the information is, for any given behavioural decision of the animal, completely irrelevant. To avoid walking into a tree you only need to know where its edges are; the fine detail of colour patterns on the trunk and leaves do not matter. To behave appropriately, therefore, an animal has to select its information, and that is exactly what it does. Some of the selection is done in the sense organ itself, which will only sense certain patterns; and the rest of the selection is carried out as the information is centrally integrated into the nervous system. Given that animals are selecting from the available information in their environment, how can we find out what they are actually responding to? One method is neurophysiology. We can record the activity of sensory neurons when an animal is successively presented with a variety of objects.

We shall meet one example of this kind of study in the prey-catching behaviour of the toad. The same kind of question can also be tackled by a grosser method. We can ignore the physiological intermediary details, and find out what kind of environmental stimuli the animal responds to at the behavioural level. The simplest kind of sensory information is used in the responses called *'kinesis'* and *'taxis'*. In a kinetic response, the animal alters its rate of movement, in a random direction, according to the intensity of the stimulus. When the stimulus, which might be light, or moisture, is of the right intensity it slows down and thus spends more of its time under those conditions. Woodlice show a kinetic response to moisture. They move faster where it is drier, and therefore spend more time where it is moist.

The unicellular organism called *Paramecium* shows a kinetic response with respect to the local concentration of carbon dioxide. A taxic response, however, is directional. Negative phototaxis, for example, means that the animal moves away from light, as in fact does the maggot of the bluebottle fly. The taxic response is made by sensing the direction of the light, which is achieved by different techniques in different species. It requires only the most elementary kind of sensory information. The animal does not need to known anything about the light source, only that it is light and where it is coming from. Paramecia normally swim about engulfing food particles at random. They feed by using the cilia around their mouth to sweep bacteria and other food particles into their food vacuoles. If a group of paramecia are placed in a drop of water, they will soon gather around and feed upon any bacterial colonies present in the drop. This action implies that their feeding is not at random. For some reason they remain quite stationary when they come upon a colony of bacteria. Other experiments have shown that they will also collect around a variety of materials–cloth, wool and cotton| Their behaviour, at least in this case, appears to be in response to a certain class of objects and is not directed towards a specific goal. We shall see further examples of this kind of behaviour in multicellular organisms.

Paramecia sometimes group together even if there are no clumps of material. How would you account for this action? Experiments have shown that they will congregate near areas that are slightly acid. Some of these experiments are illustrated. Apparently, carbon dioxide released during their own respiration upon them, survived, and continued to multiply. Those paramecia which did not possess this behavioural characteristic would not be so apt to survive and leave descendants. The type of behaviour we have just described can usually be explained in simple chemical or physical terms. It is called stereotyped behaviour a stereotype being anything that is fixed, unvarying, and showing little or no individuality. Such protist behaviour is not only stereotyped; it is also innate, or inherited. The paramecium did not learn that an acid solution means food. It positive response to an acid condition was innate. Few behaviour patterns, except perhaps chemical responses to food (or pheromones, are guided simply by the intensity of one stimulus.

More of the behaviour patterns that we can see animals performing are controlled also by the pattern of the stimulus in the environment. For example, in the case of light, the exact distribution of light and exact distribution of light and shade, which defines the shape of the image, would be important, rather than only the presence or absence of light. Animals must be able to recognize patterns in the environment, gulls do likewise. But, what exactly is it that stimulates the behaviour? How does a gull recognize an egg? Gerard Baerends and his colleagues made model eggs, which varied in their size, stippling, shape and colour. They presented herring gulls, on their nests, with choices of different model eggs, and recorded which ones were preferentially retrieved.

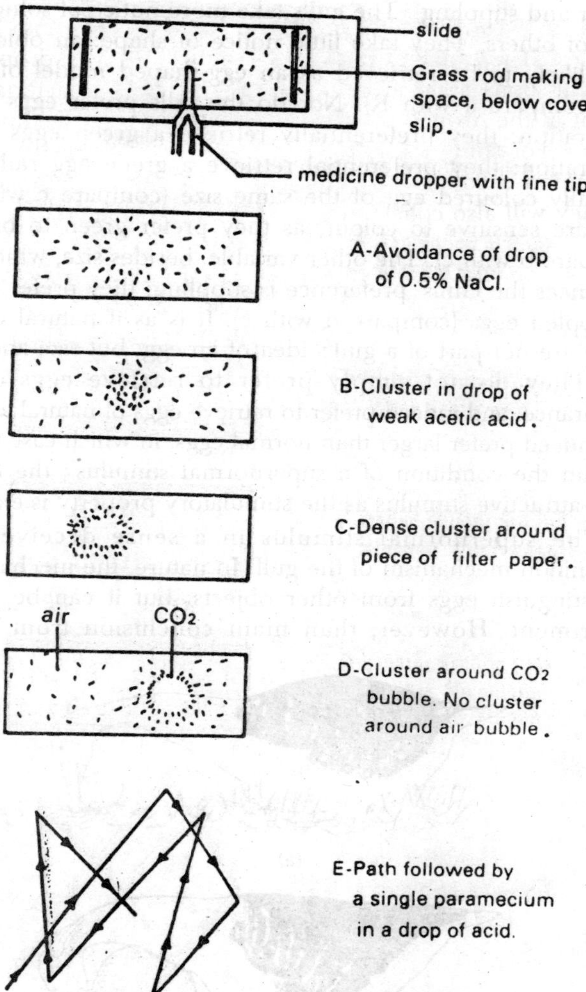


*Fig. 1.12. Behavioural experiments with the paramecium.*

The gulls preferentially retrieve larger models, even if their size exceeds the natural size of an egg, and the sizes of an egg can therefore be used as a scale of comparison for the other variables, Baerends tested each model egg by giving gulls a choice between the model egg (which might be varied in its colour, shape or stippling) and a range of size of control models (of normal shape, colour and stippling). The gulls take more notice of some variables than of others, They take little notice of shape: an oblong model is as likely to be retrieved as an egg-shaped model of the same size (compare a with R). Nor do the fulls prefer eggs of natural colouration: they preferentially retrieve a green eggs of natural colouration: they preferential retrieve a green egg rather than a naturally coloured egg of the same size (compare c with R); but they are sensitive to colour, as they prefer green to brown eggs (compare b with c). The other variable, besides size, which strongly influences the Gulls' preference is stippling; they prefer stippled to unstippled eggs (compare d with c). It is as if natural colour and shape are not part of a gull's idea of an egg, but size and stippling are. They do not simply prefer to retrieve eggs of natural appearance, and indeed prefer to retrieve eggs of natural appearance, and indeed prefer larger than normal eggs, in which case the models take on the condition of a supernormal stimulus'; the model is a more attractive stimulus as the stimulatory property is exaggerated.

The supernormal stimulus in a sense deceives the egg recognition mechanism of the gull. In nature, the mechanism work to distinguish eggs from other objects; but it can be tricked by experiment. However, than main conclusion from Baerends'

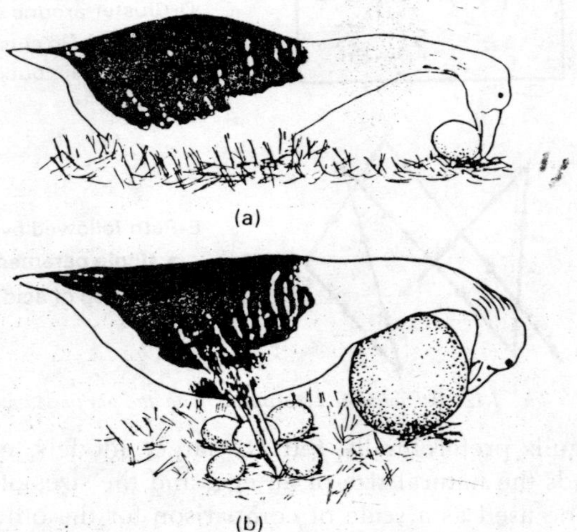


*Fig. 1.13. Greylag goose retrieving an egg which is outside the nest. This movement is very stereotyped in form and used by many ground-nesting birds. The goose attempts to retrieve a giant egg in precisely the same fashion.*

experiment is that herring gulls recognize eggs mainly by the criteria of size and stippling. '*Behavioural assays*', such as the egg retrieval response of birds, not only reveal what stimulus patterns is recognized by the animal; they are also a revelatory method of studying the sensory powers of animals. If an animal can be shown, by appropriately controlled experiments, to behave in response to some property of the environment, it must be able to sense it. Karl von Frisch applied the method to demonstrate the hearing ability and colour sensitivity of fish. In that case, the physiologist von Hess had asserted that fish are colour blind and deaf; von Frisch doubted the assertion, and he successfully trained minnows to distinguish colours by rewarding them with food, and catfish to come out of a tube when he below a whistle. In both experiments he used a behavioural response to discover a sensory ability.

## Behaviour Patterns

The behavioural output of an animal emerges as a sequence of many different behaviour patterns, and each change in the sequence can be through of as a behavioural 'choice' made by the animal. The question is how animals make those choices. Some will be responses to changed environmental stimuli. New sensory information is one factor causing an animal to choose one behaviour pattern rather than another, but it is not the only one. The same animal may not respond to the same stimulus in the same way on different occasions, and it may change what it is doing even while the environment appears to be constant. There are two, related reasons why an animal may behave differently when under similar environmental conditions. One is that its internal tendency to behave in a certain way may change. For example, when food is presented, it will become less likely to feed as it grown less hungry.

The other reason is the interaction of behavioural preferences; an animal may stop feeding in order to avoid a predator. The two reasons are related because the internal tendency of an animal to behave in a certain way is presumably determined by a balancing act among the consequences of all its possible responses to a given environment. The balancing act, and the resultant behavioural preferences or tendencies, are called the motivation of the animal. Let us consider how the motivation of an animal influences its behavioural choices. The behaviour in the freshwater fish called guppies (*Poecilia reticulata*), provides a clear example of the interaction of motivation and external stimulus.

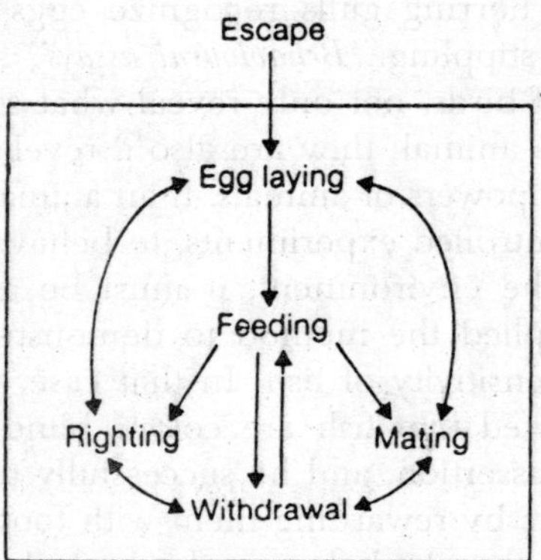


*Fig. 1.14. The behavioural priorities of the intertidal snail pleurobranchia.*

The behaviour in question is the courtship of females by males. *Baerends* recognized three different behaviour patterns that a male may perform when courting a female: 'posturing' in front of a female, limited sigmoid movement, and a full sigmoid display. How does a male decide which to perform? The answer seems to depend on the size of the female and the male's own motivation to court, which can be independently measured by his colouration. Figure depicts the combinations of these two factors necessary for a male to court a female in each of the three ways.

The particular shape of the graphs is not important here; they are only to illustrate a general point, which is that, for any behaviour pattern in any species there will be some such graph of motivational tendency and external stimulus which describes the conditions under which it is performed. The guppy illustrates choice among different behaviour patterns of one class, courtship. What of interactions among different kinds of behavioural goal? Here we need a new example, which (unlike the courtship of guppies) is understood neurophysiologically. Actually, little progress has been made in the neurophysiological study of behavioural choices. Nervous analysis is difficult enough for single behaviour units, let alone interactions among many activities.

The study of the gastropod *Pleurobranchia* by *J. W. Davis* and his colleagues has, however, partly uncovered the neurophysiological control of six behaviour patterns. The six are feeding, egg laying, escape, withdrawal of the oral veil, righting, and mating. Take first the interaction of feeding and egg laying *Pleurobranchia* is a carnivorous snail which includes eggs in its diet. When a *Pleurobranchia* lays its own eggs, it switches off its feeding habit. Another behaviour pattern, escaping from predators, is performed in preference to all other activities. A fourth behaviour patter is to withdraw its oral veil on being touched. A snail with its oral veil withdrawn cannot feed, and its tendency to withdraw its veil interacts with its tendency to feed. If food is abundant, or the snail is not hungry, withdrawal has priority over feeding, and *vice versa* when food is scarce and the snail is hungry.

The other two activities studies by *Davis* are *righting* (turning the right way up) and mating. Having established the behavioural priorities by observation, he proceeded to their neurophysiology. The system has not been completely elucidated, but it is now known for instance that two neurons are responsible for inhabiting the 'withdrawal' response when a *Pleurobranchia* is feeding, and that the inhibition of feeding during egg laying is effected hormonally. The priorities of *Pleurobranchia*, by the way, do make sense, for without them it would eat its own eggs after laying them; and if escaping from danger did not have absolute priority, it would not survive to exercise its other behavioural preferences. Hormonal influences on behaviour can be divided roughly in two categories: educational and organizational effects.

In *educational effects* hormones act as triggering influences on the expression an informance of behavioural patterns. The *organizational* effects of hormones are manifested during an organism's development. The following classification will give a clear understanding of different types of effects (in behavioural aspects).

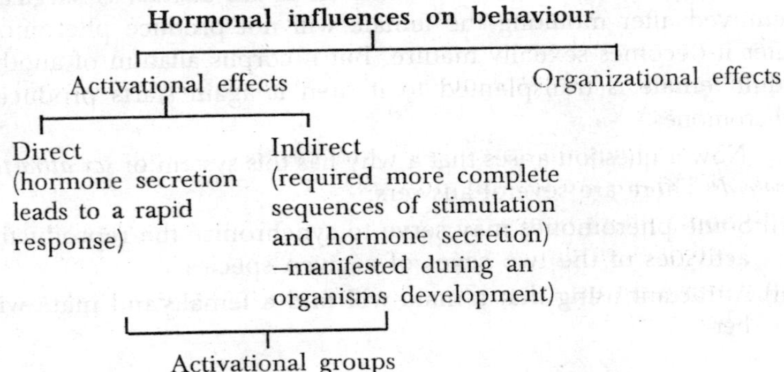

1. Sexual attraction
2. Eclosion
3. Development phrases
4. Moulting
5. Colour Change
6. Aggression and sexual behaviour
7. Secondary sexual characteristics

**Interrelationship between Hormones and Behaviour**

Investigations have used several technique to explore the between hormones and behaviour. These include:

1. *Radioimmunoassay* to directly measure circulating levels of a hormone though the use of immunological methods.
2. *Blood transfusion* to transfer the "hormonal state" of one animal at another in order to observe the behavioural effect
3. *Exterpation or removal, of a particular endocrine gland* to asses the absence of a specific hormone on behaviour.
4. *Bioassays* to indirectly assess circulating hormones levels by measuring a secondary characteristics such as skin gland, that is dependant on a particular hormone.
5. *Autoradiography* to localize the sites at which hormone uptake occurs.
6. *Hormone replacement therapy* injection of specific hormone or transplantation of a gland.

## A. Activational Effects

**(a) Sexual Attraction**

In certain species of cockroaches and moths the female release *pheromones* which act as sex attractants for male.

*Example.* If in a female cockroach *corpus allatum* is surgically removed after moulting the female will not produce pheromone after it becomes sexually mature. But if corpus allatum of another adult female is transplanted to it then it again starts producing pheromones.

Now a question arises that a why has this system of *sex attraction evolved*? There are several answers:

(i) Some pheromones may serve to synchronize the reproductive activities of the two sexes of a given species.

(ii) Attractant using that a male will find a female and mate with her.

(iii) Series as a sexual excitatory function by bringing with members logically.

(iv) Emission of a species sex- attractant may be species isolating mechanism. It avoids gamete wastage.

**(b) Eclosion**

The process whereby the adult form of an insect emerges from the pupa after metamorphosis is called *eclosion* and is another activational effect controlled hormonally. Many moth species enclose at species *specific line* of adry. The eclosion hormone which is produced by neurosecretory cells in the brain plays a critical role in this process. (Truman, 19/1; Truman and Riddifoord 1920).

*Example.* If the eclosion hormone is cirficted into pupa that are near the end of the metamorphosis eclosion behaviour, such as abdomen movements and being spreading after emergence can be activated at any time of the deny. Moths that have their brains removed usually emerge; therefore, the presence of eclosion hormone is not an absolute requirement for eclosion take place. The process, however, is not a as coordinated in brainless subjects and some activities (e.g. wing spreading )are usually absent. Thus although the hormone may not be necessary for eclosion, it does appear to be necessary for proper coordination of the sequence.

**(c) Development Phases**

In adult male desert locusts sexual behaviour is exhibited when corpora allata are removed and when corpora allata is transplanted from other locust is restores its sexual behaviour (Lohrer 1961, Pener 1965), however, similar experiments have revealed that corpora allata are not needed for sexual behaviour in certain grasshoppers.

**(d) Moulting**

In some types of norms and molluscs and uncrustaceous in investors have concentrated the presence of one or more neurosecretory or endocrine glands whose secretions effect several differentiation and malivation of gaindes and ethaulati reproduction. Many crustacean moult periodically as they grind. Removal of both eyestalk in these animals shortens the interval between, moults. If these are given the extracts of particular clearly, the gland produces a moult-inhibiting factor.

**(e) Colour Change**

In short tailed weasels which undergoes seasonal changes in pelage, orcont during spring and fall melts. In spring *metamorphose stimulating hormone* (MSH) secretion increases and new brown hairs replace white coat color during the full months. MSH secretion is inhibited by the action of another hormone *melatonin* secretion by the sexual gland, the hair then is not argumented and returns to white.

**(f) Aggression and Sexual Behaviour**

*Example.* When male ring doves are *castrated* they show decreased levels of aggressions courtship and copulation behaviour when treated with crystalline *testosterone* the normal levels of behaviours restored (*Banfield* 1971) *Mutchusin* (1969, 1971, 1978). Testosterone effects both on sexual and aggressive behaviour.

**(g) Secondary Sexual Characteristics**

*Example* (i) The characteristics cock's becomes greatly decreased in castrated roosters. (ii) Male cats, which spray urine probably as a marking behaviour, often cease to spray after their testes are removed.

## B. Organizational Effects

*Example.* Neonatal male rats pups were injected with estrogen. Histological examinations of the rats revealed some degeneration of the seminiferous tubles, where sperms are produced. The investigations mated that although these males showed mounting behaviour was irregular- the mounts were often incorrectly oriented and no ejaculation occurred. The *organizational effects* of hormones on female behaviour have been conducted on guinea pigs and Rhesus monkey. Female progeny of females that were treated with androgens when they were frequent have external genitalia that were masculinized and they exhibited male like sexual behaviour.

## Endocrine Environment Behaviour Interactions

Some activational effects involve complex interactions among behaviours, hormones and specific environmental estimate. Will discuss in detail-one example reproductive sequences in ring dives which illustrate this interrelationships.

### Reproductive Sequences in Ring Doves

At each stage on this sequence, the internal state of each kind interacts with external variables to produce the observed behaviour pattern. The variables consists of:

(i) Environmental eves, such as nests and eggs that influences hormonal and behavioural changes in both.

(ii) The behaviour of each member of the pair that stimulates changes in the hormonal levels and behaviour of its mate.

(iii) The hormonal state of both the male female dove, excluding feedback lops.

To delimine whether the presence of a mate or the nesting materials affects incubation behaviours of female ring doves. *Daniel Lehrenan* and his colleagues used three experimental groups.

(i) Females housed with a male and nesting material.

(ii) Females housed with a mate only.

(iii) Control for males housed alone.

They assessed the results of these pairings in terms of the percentages of test females in each group that exhibited incubation behaviour when presented with a nest containing eggs. Control females never incubated eggs. By days 6,7 and 8 after pairing increasing percentage of females caged with both a male and nesting material incubated test eggs. The concluded that presence of both a female and nesting materials is necessary for complete incubation behaviour in a male.

Now a question may arise that why has finely tuned system of complex interaction among behaviour, hormones and external colours evolved in ring doves. The answer to this question is that since reproductive cycle involves the dual partical ration of both and nececlatis that their activities be coordinated throughout the reproductive success of the pain is guaranteed only if both perform certain acts in syutony e.g. Females that laid eggs before a nest was completed would contributed little to future generations. If the male developed a crop or began to produce crop milk before any rquats had happened etc.

**Migratory Behaviour in Birds**

Migration in behaviour in birds is also due to the result of the interaction between the environments and the hormones. *Reowan* observed that this peak of gonadal and the recurdiocence coincides with the peak of migration urge. Rowan's suggested that many factors influences migration. Increasing day length increasing ambicint temperature other environmental changes which affect the animal through pituitary activity and perhaps the nervous system directly.

## MECHANISMS AND ONTOGENY OF HORMONAL EFFECTS

**Stimuli and Mechanisms**

To understand the actions of a hormone, both the stimulation and its release and its effects upon behaviour must be known. The mechanisms by which hormones effects behaviour through their effects upon.

(i) The whole organism (e.g. general activity level).
(ii) Morphologic structure employed in specific response patterns.
(iii) Peripheral receptor mechanisms.
(iv) Integrative functions of the C.N.S.
  (a) Control of this development of nervous organisation.
  (b) Control periodic growth and regression of nervous elements.
  (c) Control of sensitivity of stimulation.

**Ontogenetic Development**

If baby which is injected with testosterone, complex mating responses including covering treading have been observed. There are many examples which ultimately conclude that some interaction of hormonal action and behavioural development establishes a behavioural organization which then persists in the absence of hormonal influences.

**Some Recent Developments**

Since 1950 research has centered increasingly upon the problem of untangling the delicate and complex interrelationships between hormones, the nervous system and behaviour.

(i) In some what similar program of research *Hinde* (e.g. 1985) has been exploring the hormonal bases and external stimulus conditions, controlling nest building.
(ii) Active research continues to the field of migration too. The *Wolfsen theory* stated the accumulation of fat sets off migratory behaviour has been challenged by the schools.
(iii) Another interesting approach to hormones and behaviour has been to emplified by the work of *R.P. Michale* regarding sexuality in the verge of *nymphonania.*
(iv) *D.S. Lehrman* (e.g. 1964) has recently made three discoveries the role of prolactin in parental behaviour in doves.

*First* prolactin alone does not elicit parental behaviour, but requires estrogenic *priming.*

*Second* low or hormone can act directly on a peripheral structure involved in behaviour. Without involving vagus effects on the C.N.S.

*Third* The young birds provide the tactile stimuli that elicit regurgitation from the crop.

*Farmer* and *King* have shown by ingenious experiments that if captive birds were not allowed to accumulate far *Zugunrube* appears indepently to response to long photoperiod.

# 2

# Instinct Behaviour

Almost all animals are composed of the same basic materials, the difference between species resulting from differences in the way in which these basic materials are put together. Although this regulatory function is poorly understood, it is known that genes control development by producing proteins that regulate the complex organization of embryological progress. How does an animal's behaviour become so well fitted to its normal environment? There are two basic ways. Firstly, it may be born with the right responses 'built in' to the nervous system as part of its inherited structure. Honey-bees inherit the ability to form wings and wing muscles for flight; they also inherit the tendency to fly towards flowers and seek nectar and pollen. Such responses are popularly called 'instinctive'–a term which has often been abused bur remains useful.

Instinctive behaviour evolves gradually, as do structural features, and natural selection modifies it to fit the environment in the best way. It forms a kind of' species memory' passed on from each generation to its offspring's. Alternatively an animal may have no inherited responsiveness with regard to a situation but instead have the ability to modify its behaviour in the light of experience. It learns which responses give the best results and changes its behaviour accordingly. Instinct and learning both ensure adaptive behaviour, the former by selection operating during the history of a species, the latter during the history of an individual. Stated in this way there is a clear dichotomy, which is realistic when actual examples are examined. But before we do this it is worth considering the importance of instinct and learning in a general way through the animal kingdom.

## CONDITIONED REFLEX

*Watson* studied learning in animals such as rats and was impressed by the flexibility of their behaviour and by how learning could lead them to behave in an adaptive manner. He believed that they built up connections in their brains as a result of perceiving associations in the outside world, so they would associate a particular place with food and run through a maze to get there. His views were partly stimulated by the work of *Ivan Pavlov*, the Russian physiologist, who had studied conditioned reflexes in dogs. *Pavlov* showed how a new stimulus could be made to elicit a reflex as a result of the animal building up as association. In his most famous experiment, a dog was presented with food, a sti. ulus which normally makes dogs produce saliva in readiness to eat, and at the same time a bell was rung. This normally has no effect on salivation. After a number of such presentations, the bell was suddenly sounded without any food being produced the dog salivated.

The animal had formed an association between the bell and the food so that both were now capable of producing the response. This sort of training is referred to as conditioning and the result, in this case salivation in response to a bell, as a conditioned reflex. The main stress in the theories of *Pavlov and Watson* was not on rewards : the dog did not need to eat the food to form the connection. Later ideas have suggested that a great deal of learning does rely on reward, the most notable exponent of this view being *B.F Skinner*. Like *Watson* before him. *Skinner* believer that the great

*Fig. 2.1. Animal in a set-up of a Pavlovian study of conditioned responses.*

majority of behaviour, with the exception of simple reflexes, results from experience. But he stresses the way in which the responses of animals come to be linked to particular stimuli as a result of reward. Animals tend to repeat rewarded actions and reward can thus be used to alter their behaviour.

The dove in the Skinner box referred to in the last chapter soon learns that pecking the right-hand key gives it water while the left-hand key yields food. Based on ideas such as these, elaborate theories of animal learning have been built up which stress the role of the environment and of experience, rather than genetics and inheritance, in the development of behaviour. At an extreme, such psychologists have suggested that learning in all animals is subject to the same rules and that animals do not have to have predisposition's about the sort of things they will learn but will build up any associations with equal ease provided that their sensory and motor systems allow them to do so.

Obviously one would not attempt to train a sparrow to fly in the dark or a tortoise to walk on its hind legs. The following famous assertion of *Watson's* illustrates the fervour of his belief that the environment rather than heredity was the major determinant of behaviour : 'Give me a dozen healthy infants, well-formed, and my own specified world to bring them up in and I'll guarantee to take any one at random and train him to become any type of specialist I might suggest doctor, lawyer, artist, merchant-chief and yes, even beggar man and thief, regardless of his ancestors.' He went on to admit that he was over stating the case, but insisted he was doing so as a counter to the sweeping claims of his opponents.

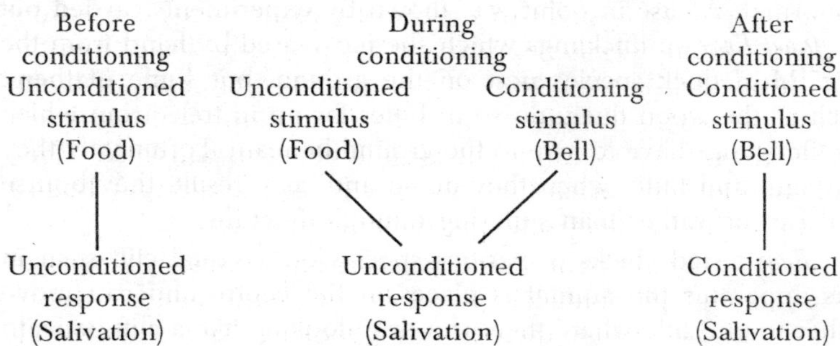


*Fig. 2.2. Conditioning involves the formation of new connections.*

Both views such as this and those of the ethnologists could not be right, but to some extent they existed because their adherents were studying different things. Psychologists were interested in learning and in behaviour patterns, like the lever pressing of a rat in a Skinner box, which could be modified by it so leading an animal to behave differently from other members of its species. By contrast, the ethnologists were interested in more fixed behaviour, such as courtship displays, which were common to all members of a species. Why did they consider such a behaviour to be 'innate' and how does the evidence they used hold up to detailed examination?

**Ethnologists and Instinct**

In 1828 a boy of about 16 was found in the market-place at *Nuremberg* in Germany. He could not speak but his behaviour was like that of a child, so he was labelled *the wild boy.* His name was *Kaspar Hauser* and, when he was able to communicate, he explained that he had been brought up entirely in isolation by a man who had kept him and cared for him all by himself in a hole. *Kaspar Hauser* may well have been a fraud, but he has given his name to the deprivation experiments which have often been used by ethnologists in an attempt to understand whether or not the behaviour they were studying was innate.

The idea was to deprive an animal of relevant aspects of experience and, if the behaviour appeared despite this, it could be regarded as inherited whereas, if it did not appear or developed abnormally, it could be assumed that the missing experience was important. There are many striking examples of behaviour appearing apparently normally in animals whose experience is distinctly abnormal. A case in point was shown by experiments carried out by *Janet Kear* on ducklings which she has reared by hand from the egg. Most duck species nest on the ground, but some of them, such as the wood duck, do so in holes far up in trees from which the fledglings have to leap to the ground beneath. Fortunately they are light and fluffy when they do so and, as a result, they bourse and run off rather than suffering multiple fractures.

*Kear* tested chicks of various species on a visual cliff such in this apparatus the animal is places in the centre and can move either to one side where there is a drop looking like a cliff beneath the glass or to the other where the floor is immediately under the glass so that it looks shallow. The behaviour of the ducklings was appropriate to their normal nesting place. Tree-nesters did not avoid the deep side of the cliff but, if they moved that way, they would

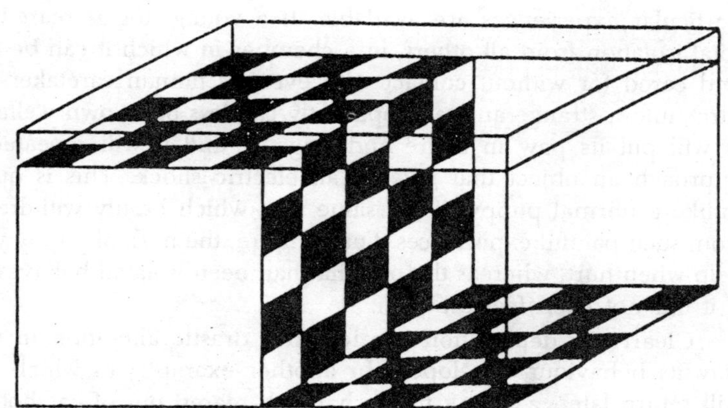

*Fig. 2.3. A visual cliff, as used by Kear in her studies of ducklings. The whole apparatus is covered by a sheet of glass, but the left side appear to be shallow while that on the right looks deep.*

leap as if casting themselves into space. On the other hand, the ground-nesters tended to move to the shallow side rather that the deep one, suggesting that they avoided heights. Interestingly, if they did move to the deep side their behaviour was quite different they pushed off with both feet as they would when moving out from the edge of a pond! Being hand-reared, these young birds had and no opportunity to learn the actions they showed from others or from earlier experience with ponds or with cliffs.

Many ethologists would therefore have used this evidence to argue that the behaviour must be innate because it develops despite deprivation of opportunities for learning, and the main motivation for carrying out such experiments has often been to discover whether behaviour is *innate* or *learnt.* Sometimes, as with the ducklings, behaviour develops normally even though the animal is reared in a very impoverished environment. Another good example here is the hoarding behaviour of squirrels whereby, even in captivity, they will bury nuts underground to form stores which they eat later when food is scarce.

If such an animal is raised on a liquid diet, so that it never experiences nuts, with masses of this food available the whole time, so that it never needs to hoard, and on a bare floor so that digging is impossible, the first time it encounters nuts and earth it still digs a hole and buries them. By contrast with these experiments, others have shown behaviour patterns to be radically altered unless particular experiences are available. If a young dog is reared in total isolation from all others, in a chamber in which it can be fed and cared for without contact with even its human caretakers, it turns into a strange animal, apparently careless of its own welfare. It will put its paw in a fire and singe it, and it will repeatedly approach an object that gives it an electric shock. This is quite unlike a normal puppy of the same age, which hastily withdraws from such painful experiences. Furthermore, the normal puppy will yelp when hurt, whereas the one that had been isolated behaves as if it did not even feel the pain.

Clearly its deprivation has led to a drastic alteration in the way its behaviour developed. In another example, to which we will return later, a young male chaffinch reared out of earshot of all birds of its species has been found to develop a very simple and unstructured song quite unlike that of a normal adult. If he is deafened as well, so that he cannot even hear his own efforts at singing, the song he produces is even worse, bring little more than

*Fig. 2.4. Studies of behavioural development in dogs involved rearing puppies in isolation both from other dogs and from humans. They were kept in chambers with two doors and a partition so that one part could be cleaned out while they were in other. As a result, they were isolated from all contact with the outside.*

a screech. These *Kasper Hauser* experiments vary enormously in what they actually deny the animals. The isolated chaffinch cannot copy song from other birds, but he can practice singing; if deafened he can still practice, but cannot here the outcome.

The hoarding squirrel has had experience of neither nuts nor earth so its deprivation is more extreme: it cannot copy from others, it cannot learn for itself, nor can it practice digging in earth, though it can carry out the movements concerned on the bare cage floor. In many cases it has proved very difficult to deny animals all the experiences one might think likely to be relevant. Finches with no hay that they can use for nest building will carry seed, lettuce, faeces and feathers to their nest site. Some bizarre behaviour results when the feathers used are still attached to themselves or to their mates. A bird may have to walk rather than fly to its nest site because it is holding its own wing in it beak as nest material; after carefully placing the wing in a corner of the nest it will then fly down, pick the wing up again and struggle back to the nest with it!

## The Characteristics of Instinct and Learning

We must now examine more closely the characteristics of instinct and learning. There are two conspicuous features of instinctive behaviour which may seem, at first sight, to be unique. First, that it consists of rigid, stereotyped patterns of movement which are very similar in all individuals of a species; all differ- wasps of the same species build their nests in the same way, domestic cockerels all use the same series of movements when courting hens, and so on. Secondly, instinctive patterns can often be evoked most readily by very simple stimuli. When presented with a complex situation the animal responds to one part of it and virtually ignores the rest. A robin displays more aggression to a tuft red feathers from the breast of a rival male than to a complete bird which lacks only these feathers. However, striking, such characteristics are quite inadequate to distinguish instinctive from learnt behaviour.

The latter is often described rather vaguely as being 'more flexible', but in fact the patterns of movement involved may be just as stereotyped as those of instinct. Rates placed in a box where they must learn to press on a projecting lever on order to obtain a pellet of food will develop a particular manner of doing so. Some use always their left paw, others press with their chins, and

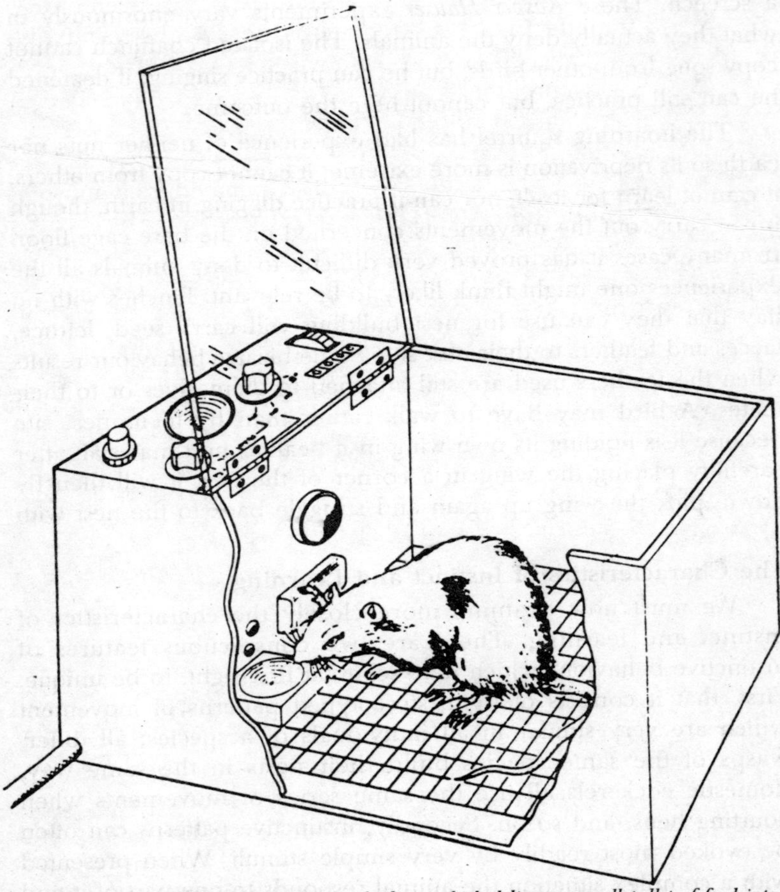

*Fig. 2.5. Rat in a Skinner box pressing the lever which delivers a small pellet of food into the cup.*

individual rats tend to be conservative in the method they use. In his delightful book, *King Solomon's Ring, Lorenz* describes how water shrews learn the geography of their environment in amazing detail. If at one point on a trail they have to jump over a small log, this movement is learnt with such fixity that they continue to make the jump in precisely the same fashion long after the obstacle is removed.

There are plenty of other examples where animals have comparable difficulty in 'unlearning' something. The movements appear to become almost 'automatic' and may persist even of they are no longer effective. Learnt patterns nearly always involve responding to particular cues in the environment and, just as with instinctive behaviour, other features may be ignored. It is relatively easy to train animals to discriminate one key stimulus within a complex changing situation. If we are to make any critical distinction between instinctive and learnt behaviour; it must be based not upon their overt characteristics but upon their development within the individual. Rats learn to press a food lever and each does so in a stereotyped way, but this way is unique to the rat and varies between rats. All sexually receptive female rats show the same stereotyped acceptance posture when mounted by a male.

*Fig. 2.6. Copulation in rats. The female's posture with raised pelvis and deflected tail is very stereotyped in form. The pressure of the male's fore-limbs on her flanks is one of the stimuli necessary for the female to respond.*

Furthermore, if we put a hungry rat into a box with a food lever, it may be hours before, by chance, it pushes against the lever. But if a virgin female rat is brought into receptive condition by a hormone injection, it assumes the acceptance posture after very little experience with a male. Here we have the basis for developmental criteria of instinctive behaviour and the commonest way of testing them has been the so-called 'isolation experiment'. Animals are kept individually out of contact with others from as early an age as possible. When mature, their responses to a variety of stimuli are tested and compared with those of animals reared normally. Rather few animals have been tested under really rigorous conditions but some fish and birds have been shown to perform various feeding, sexual and alarm patterns of behaviour quite normally after being reared in isolation.

The life histories of many insects are natural isolation experiments which demonstrate vividly that no practice or learning are involved in the behavioural development of the adults. However, easily isolation experiments enable us to eliminate the possibility that animals learn how to do something, they offer only a very restricted view of the factors that may be operating during development–learning and practice are only two from a wide range of possibilities. The point is that isolation can tell us only what factors are *not* important for the development of behaviour, it tells us nothing of what is involved and this may not be at all obvious. For example, young mallard ducklings hatched from an incubator respond preferentially to the call notes of mallard ducks over other related calls the first time that they hear them. They clearly show an instinctive preference for their own species call. However, Gottlieb 171-2 has shown that one factor which contributes to the development of this preference is the embryo duckling's ability to here the calls–quite unlike those of the mother–that itself and other ducklings make whilst still inside the eggshell. Nobody would doubt that the development of many behaviour patterns must be under genetic control and result from an inherited potentiality of the central nervous system. But this is the point from which studies of development should take off. We should certainly must be content of label such patterns as 'instinctive' and leave it at that. Genes may control behavioural development, but to do so they must interact with the developing animals' environment.

Gottlieb's ducklings need some auditory stimulation if they are to be primed to respond to the maternal calls. It will be our task to discover such factors in the behavioural environment of animals which affect the development of their instinctive patterns. Clearly we must avoid any simple dichotomy which ascribes instinct to the genes and learning to the environment. Both must be involved in the development of all behaviour. This point might not seem to need such emphasis, but in fact misunderstandings here have been, until quite recently, the basis for a considerable dispute among animal behaviour workers.

## Development of Behaviour

May studies of the development of behaviour are designed to investigate the mechanisms that underlie behaviour patterns observed in adult animals. Such a study may be relatively direct–for example, determining what factor affect nest-sit selection by a bird species–or it may be a complex investigation–for example, establishing the ontogenetic determinants of feeding strategies in omnivorous rodents. Some studies apply conditions of either deprivation or enrichment; others alter the quality of specific types of stimulation provided during ontogeny rather than vary the overall quantity of stimulation. There types of investigations provide useful information on the development of traits within natural ecological and social context.

Other studies of behaviour development utilize a variety of manipulations to test the effects of various treatments. Examples of these approaches are presented when we examine the chronology of development. First, a bit of elaboration is needed on some aspects of experimental design and testing procedures used by investigators studying the development of animal behaviour.

**Aspects of Experimental Design**

*King* set forth seven parameters relevant to the study of early experience. His scheme, or portions of it, can be rather broadly used in designing many of the experiments in behaviour development–particularly those that attempt to assess the effects of early experience on behaviour.

Parameters of the early experience treatment:

1. Age of the animal at the time the early experience treatment is given.
2. Type or quality of the early experience treatment.
3. Duration or quality of the early experience treatment.

Parameters of the tests administered to determine the effects of early experience treatment:

4. Age of the animal at the time of testing
5. Type of test used to assess the effects of early experience treatment.
6. Testing for the persistence of the early experience treatment.

The parameter of the genetics of the species being investigated.

7. Testing different strains or species to determine the effects of early experience on different animals.

**Testing Procedures**

We can test the long-term influence or persistence of early experience treatments by two types of experimental design. One method provides an animal with early experience treatment and tests for the effects of that experience at various ages; we use the same animal for each test. This type is called *longitudinal design* and requires the use of appropriate repeated-measures statistical tests to analyse the results. Alternatively, we can provide an animal with an early experience treatment and test it only once at a later age, because in a second test of the same animal we would not be able to tell whether the observed effect was due to the treatment or to the experience of the first test.

We therefore avoid re-using the test subject, but possibly we lose some information about the persistence of early experience treatment on that subject. This second type of design is called *cross-sectional.* In some studies of early experience, investigators subject animals to stimulus deprivation or enrichment by manipulating either the quality or quantity of one of three types of stimulation or some combination : (1) sensory, (2) motor, or (3) social.

In every case investigations compare the test performance of animals reared under deprived or enriched conditions with the test performance of control subjects reared under *normal conditions. Normal rearing* in this context usually refers to the common laboratory environment–a cage that restricts the amount and type of activity of the animal, contrived social contacts, a prepared diet, and constant ambient conditions. These conditions are extremely simplified when compared with natural environment. Thus, we should keep in mind that the control conditions are often stimulus-poor and not at all comparable with the subject's natural surroundings. In the chronological sequence which follows we provide some general remarks and several examples of investigations of behaviour development at each stage or period.

## Feeding and Food Preferences

Is it possible that the food ingested by a lactating female rat influences the later food preferences of her pups ? To test this question Galef and *Henderson* experimented with thirty-two pups born to four female rats. After the birth of the pups, the investigators removed the lactating mothers from their home cages and placed them in separate compartments during three one-hour feeding periods each day. They gave two of the females food prepared by Purina and two food made by Turtox. When untested rats are given a choice test they normally prefer the Turtox diet; the foods differ in taste, texture and colour. *Galef and Henderson* measured the food preferences of the pups for seven days, beginning at seventeen days of age. Figure shows that the young rats reared by lactating mothers fed the Purina food ate proportionately more Purina chow than Turtox; and pups reared by lactating mothers fed Turtox preferred Turtox chow. A rat pup may acquire information about the food its mother is consuming in three possible ways.

Young rats may consume faeces dropped by the mother; they may ingest or smell particles of food adhering to the mother's fur or oral region; or the favour or odour of the mother's food may be transmitted in her milk. Further experiments conducted by *Galef and Henderson* have confirmed that information about diet is transmitted through the mother's milk. The result of these experiments have some interesting implications for our understanding of the development of food habits in young rodents as they leave the natal home site and fend for themselves.

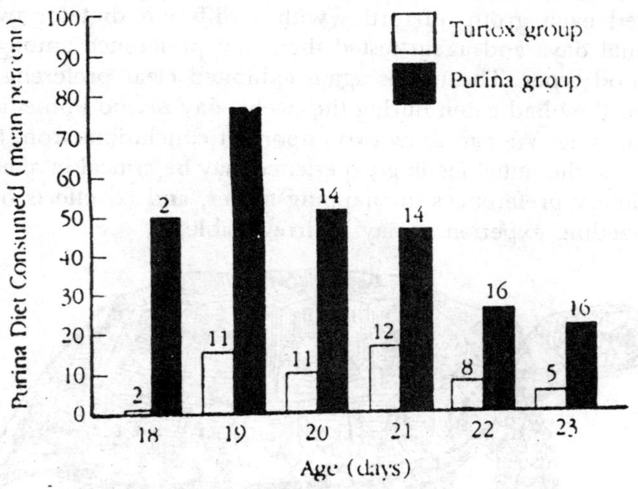


*Fig. 2.7. Feeding preferences in rat pups.*

Prior to this work the assumption had been that young rats had to learn to locate solid foods without any assistance from adult rats. However, the study outlined here and other work by *Galef* and his colleagues suggest that young rodents may in fact obtain some of this information via their mothers. At the age of weaning the feeding site selection of young rats is influenced by interactions with adults, and in particular by olfactory cues associated with conspecifics. When presented with a choice, rat pups select a feeding site associated with either conspecifics or their excreta in preference to a clean site. Interestingly, if pups are reared without contact with conspecifics, they select a feeding site without respect to whether it is clean or has conspecific cues present. When pups are reared away from conspecifics, but then are given five days exposure to conspecifics prior to testing, that exposure is sufficient for pups to show a strong preference for feeding sites associated with conspecifics stimuli.

**Food Preferences in Turtle**

*Burghardt and Hess* investigated the feeding preferences of snapping turtles (*Chelydra serpentia*). They fed separate groups of newly hatched turtles either horsemeat, fish, or worms for a twelve-day period. When they tested the turtles on day 13, with a choice of three diets, the turtles demonstrated significant preferences for the food on which they had been reared. The investigation then provided each group of turtles with a different diet for twelve additional days and again tested their diet preference among the three food types. The turtles again exhibited clear preference for the food they had eaten during the twelve-day period immediately after hatching. We can draw two important conclusions from these results : (1) the initial feeding experience may be critical in affecting later dietary preferences in snapping turtles, and (2) effects of the initial feeding experience may be irreversible.

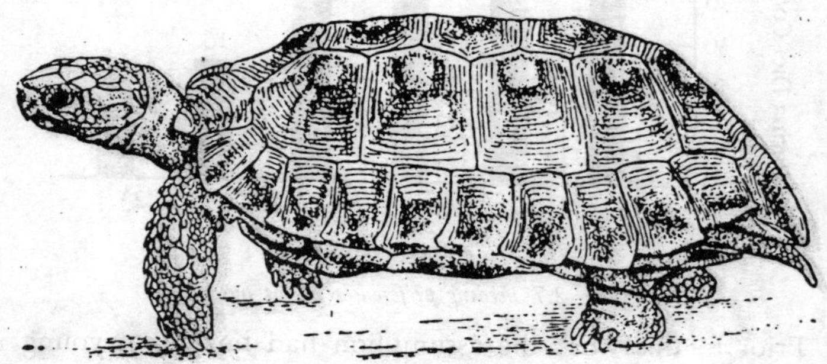

*Fig. 2.8. Testudo (land tortoise).*

Further testing would be needed to confirm and extent these conclusions to the retention into adulthood of the effects of the initial feeding experience, but some type of food imprinting may be occurring–that is, the turtle's food preference may have become fixed even through the turtle with still consume other foods.

**Feeding Behaviour of Gull Chicks**

The feeding behaviour of laughing gull (*Larus atricilla*) has been studied in both field and laboratory settings. *Tinbergen* and *Perdeck* made observations in the field of the effects of holding cardboard models in front of young gull chicks in their nests. They found that hungry chicks use their bills in a pecking and stroking pattern directed at the bill of a parent, a behaviour that induces the adult bird to regurgitate food for the chick. *Hailman* examined how this food-begging behaviour in gull chicks developed. His results indicate that the behaviour resulted from an interaction of genes and environment.

**Table 2.1. Food Preferences of Snapping Turtles After one Meal of one of Two Foods**

| | | *First Meal* | | *Second Meal* | | *Choice Test* |
|---|---|---|---|---|---|---|
| *Group* | *N* | *Food* | *Total Pieces Eaten* | *Total Pieces Food* | *Eaten* | *Number Preferring First-Fed Food* |
| 1 | 12 | Horsemeat | 26 | Worm meat | 19 | 12 |
| 2 | 13 | Worm meat | 28 | Horsemeat | 34 | 8 |
| Totals | 25 | | 54 | | 53 | 20 |

*Fig. 2.9. Normal feeding behaviour of laughing gull chick.*

The genome of the young gull provides the information necessary for the correct maturation of the bird's sensory and motor systems. As it matures the bird learns to peck at the parent's red bill, which contrasts with its black head. By using cardboard models of a gull's head to test pecking behaviour, *Hailman* found that several days were required to attain a record of 75 to 90 per cent hits. Thus, full development of the begging behaviour requires the genetically programmed development of sensory-neural systems to receive and interpret the environmental stimuli, motor capacities for responding, and the experience associated with learning to peck for the food.

## Influence of Environment

Some forms of behaviour do not appear until a particular stage of development is reached. Some of these behaviours seem to develop without any obvious practice. For example, pigeons start to flap their wings and fly erratically at a particular age, and their flying ability appears to improve with practice. However, *Grohmann*, in a classical experiment, reared a group of pigeons in tubes so that they could not move their wings. Another group of the same age was allowed to develop without restraint. When the unrestrained pigeons had reached the stage at which they could fly satisfactorily, those that had been restrained were freed. *Grohmann* discovered that they also were able to fly immediately upon being released. Chickens that are featherless as a result of mutation develop normal wingflapping vestibular reflexes, even though there are completely ineffective.

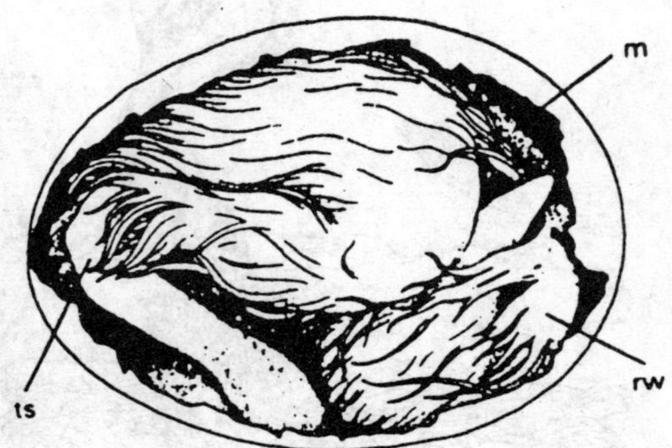


*Fig. 2.10. The hatching position of the chick embryo approximately 1-2 days before emergence from the shell, m membrane; is tarsal joint of leg right wing.*

At first sight it would appear that the behaviour is independent of environmental factors, but we must not assume that the ability to fly develops irrespective of any ontogenetic eventuality. Rather, flying ability is contingent upon some essential conditions, although our knowledge of these id sketchy. A similar example is the vocalisations of pigeons and domestic fowl, which are highly stereotyped and appear at certain stages of development. They are not dependent upon auditory experience but upon certain hormonal conditions that are a normal part of overall maturation. Many movement patterns appear to develop in the absence of practice or example. However, study of the developmental history of some behaviour patterns has suggested that they involve fragmentary and incomplete movements that might influence the course of development.

As an example, we can consider *Kuo's* (1932) work on the development of behaviour in chick embryos. *Kuo* placed windows in the eggshell and was able to directly observe the behaviour of the embryo. He found that the embryo is constantly subject to stimulation both from its own activity and from outside the egg. Various movements can be observed, including initial passive movements of the head caused by the beating heart. Active head movements appear within a few days, and these may be accompanied by opening and closing of the beak. By the seventeenth day, movements similar to pecking are apparent. Some scientists believe that the chick learns to peck in this way. Others ridicule this idea. Whether or not true learning occurs, there id little doubt that the developing embryo responds to stimuli from both inside and outside the egg and that these stimuli may influence development. Thus, *Margaret Vince* showed that in some species, stimuli produced by the embryo can influence the development of other eggs in the clutch and can lead to synchronisation of hatching.

*Gilbert Gottlieb* shows that duck embryos can respond to maternal calls five days, before hatching and that such experience may influence the duckling's subsequent behaviour. However, other workers provide evidence that movement in the chick embryo develops in the absence of sensory stimulation and that the movement patterns that occur later in ontogeny are not necessarily influenced by earlier movements. Readers should remember that the young of different species develop in very different ways. While ducklings and goat kids are mobile as soon as they are hatched or born, blackbird nestlings and kittens are helpless at this stage. We are not surprised to discover that the postnatal development of the latter is influenced by experience. Should we be surprised at prenatal influences in the case of animals that are still inside the egg or womb at the same overall stage of development? There seems little difference in principle between the protection the duckling receives while in its egg and that the young blackbird enjoys in the nest.

Similarly, there would seem to be little difference in principle between the effects of experience in the two cases. For each species, certain types of experience are essential for normal development for example, adult cat vision is abnormal if there has been any disruption in the co-ordination of the vision from its two eyes during the first few months of life–by inducing an artificial squint or by covering each eye on alternate days so that the two eyes never work together. The exact nature of the experience needed for normal development varies considerable from one species to another.

## Learning during Development

Some animals appear to be pre-programmed to learn about certain aspects of the environment during particular periods of their development, for example, language learning in humans. In contrast to other types of learning, language learning is *pre-programmed* in the sense that in the absence of any obvious reward or punishment. For example, the young of many precocial species show a fairly indiscriminate attachment to moving objects. Thus, newly hatched mallard ducklings, separated from their mother, will follow a crude model duck, a person, or even a simple box moved slowly away from them. Some stimuli are more effective that others in eliciting this following response.

In the natural environment the most effective stimuli normally are provided by the mother, and approaches to the mother often are rewarded by body contact and warmth or by food that the mother uncovers. In the laboratory the attachment to a model can be enhanced by food rewards. The more an animal develops an attachment to one object, the less interested it is in others. This process of learning, through which attachment to the mother normally develops, is called *imprinting.* It occurs during a particular, sensitive period of development, which varies according to the species and the circumstances.

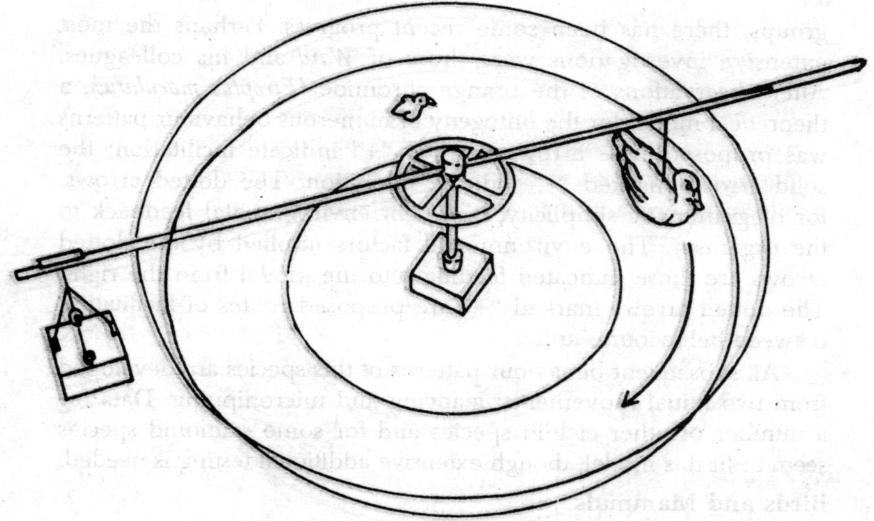

*Fig. 2.11. A duckling following its 'mother' in an apparatus designed to test aspects of imprinting.*

Imprinting may have long-term effects, beyond the attachment to a parent or foster parent. In many mammals such early experience affects subsequent social adjustment. In a number of bird species, imprinting has been shown to affect subsequent sexual behaviour. In short, imprinting is a process of learning that occurs at a particular stage of development and that affects subsequent behaviour toward parents, peers, or sexual partners. If the object of attachment is a cardboard box, then the duckling will become attached to the box as to a parent. If male zebra finches (*Taeniopygia guttata)* are raised by Bengalese finches (*Lonchura striate*), then they court Bengalese finch females when adult. Although this type off learning may be influenced by rewards, it is not dependent upon them or upon any particular consequences of the behaviour. The learning is pre-programmed to take place as part of the normal process of development and in whatever circumstances pertain at the time.

## Juvenile Events

### In Fishes

While considerable fewer studies of the ontogeny of fish behaviour have been conducted than for some other vertebrate groups, there has been some recent progress. Perhaps the most extensive investig.tions were those of *Ward* and his colleagues. After observations cf the orange chromide (*Etroplus maculatus*), a theoretical model for the ontogeny of numerous behaviour patterns was proposed. The arrows marked "+" indicate facilitation; the solid arrows marked "–" indicate inhibition. The dotted arrows, for diagrammatic simplicity, represent environmental feedback to the organism. The environmental factors implied by the dotted arrows are those indicated feeding into the model from the right. The dotted arrows marked "+" are proposed routes of facilitation between behavioural units.

All subsequent behaviour patterns of this species are developed from two initial movements; glancing and micronipping. Data for a number of other cichlid species and for some salmonid species seem to fit this model, though extensive additional testing is needed.

**Birds and Mammals**

The period which lasts roughly from fledging (birds) or weaning (mammals) until the animal is fully independent and on the way to maturity can be loosely termed the *juvenile stage*. Development is a continuous process, usually marked by various identifiable events, such as birth or fledging,. As investigators we often divide the continuum into stages or periods for convenience. Because animals of different species–and indeed individuals of the same species–often proceed through development at varying rates, the timing and length of these periods will vary. Several pivotal events usually occur during the juvenile period as we define it here, including dispersal from the natal site, puberty, learning appropriate communications signals, various experiences which influence later behaviour patterns, and, for birds, the first migration toward the equator. Play behaviour, which for some species constitutes an important activity during the juvenile period, will be discussed in more detail later in the chapter.

**Social Deprivation in Domestic Animals**

One method of assessing the possible significance of a particular experience or stimulation during development involves depriving the organism of one or more of these experiences or stimuli and measuring the effects on subsequent behaviour. Some of these treatments may actually commence prior to the juvenile period, whereas others start during this stage. The effects may be noticeable and measure-able during this stage or later in the life of the organism. Deprivation may be social, sensory, or motor, or some combination of these.

*Fuller* investigated the effects of experiential deprivation on the social behaviour of dogs (*Canis familiaris*). He and his colleagues placed beagle and terrier puppies in isolation for varying lengths of time. They included periodic contact with a human handler or another dog at prescribed intervals in some of the isolation regimes. They then gave the dogs an open-field test and several learning tasks, and observed the subjects for periods when the dogs could make contact with a towel, a ball, or another puppy. The results were similar to the findings for monkeys reported by *Harlow* and his co-workers the more severe the deprivation, the more pronounce the dog's behaviour deficits and abnormalities.

The dogs reared under the most restricted regimes sometimes failed even to leave the starting area of an open-field test, were generally less active than dogs from other treatments, and lost most frequently in competition with another puppy. Prior to Fuller's work two explanations of the observed behaviour differences between deprivation-reared and normal animals had been proposed. Some investigators ascribe the differences to the deterioration of previously organised neural patterns; this theory is sometimes referred to as the *disuse hypothesis.*

Others postulate that the behaviour deficits resulted from a lack of critical information input necessary for proper development of neural patterns and normal behavioural responses; this concept is often called the *loss-of-information hypothesis.* Fuller proposed a third explanation, which he termed the *stress-of-emergence hypothesis.* Animals that have been in isolation and that are suddenly thrust into a test situation loaded with many novel stimuli may suffer from a stimulus overload–that is, they are faced with a multitude of competing emotional responses.

To test his hypothesis *Fuller* suggested three alternative treatments : (1) providing the isolate-reared dogs several brief, pretest exposures to the testing arena to habituate them to the situation; (2) giving the dogs a tranquilisation drug, like chlorpromazine, before testing to reduce the effect of stimulus bombardment; or (3) giving the dog several brief periods of handling by the experimenter before testing. These methods have been tried and, with varying degrees of success, have produced some reductions in the stress of emergence. Fuller's hypothesis appears to be valid in some experimental deprivation studies, but we must consider the two earlier hypotheses as primary explanations for the behavioural deficits that result from isolation treatments.

**In Monkeys**

From the 1950s onward *Harlow* and his colleagues have conducted a series of studies on the effects of social deprivation on the behaviour of young, adolescent, and adult rhesus monkeys (*Macaca mulatta*). The investigators usually reared young rhesus monkeys under various conditions for periods of time during the first two years of life. The conditions included: (1) Rearing in total isolation in chambers that remove the infant from all social contacts and most external stimulation. We should note two additional points in connection with these studies of social deprivation. First, some of the behaviours that *Harlow* and others have recorded as *"abnormal"* occur occasionally in rhesus monkeys reared under "normal" conditions: self-directed aggression has been observed on a number of occasions in nonisolate-reared subjects.

Another study reported that in free-ranging rhesus monkeys– or those in their natural habitat, but with some spatial restrictions– only 50 per cent of the young born to primiparous females survive to the age of twelve months. *Harlow* and his co-workers have also noted that "motherless mothers" exhibit much better maternal behaviour with their second infant. Part of the deficiency in maternal care recorded in *"motherless mothers"* may result from their need to learn how to be good mothers and not strictly to the emotional abnormalities associated with isolate rearing. Second, and quite significantly, *Harlow* and his associated have succeeded in socially rehabilitating isolate-reared monkeys. They accomplished this by exposing six-month-old social isolates to three-month-old normal monkeys, called *therapy monkeys*, for two hours per day, three days per week, for one month. The effects produced by some types of early social isolation are not, as was once thought, irreversible, though even rehabilitated monkeys continue to exhibit some behaviour defects.

**Singing in Birds**

In the past several decades an increasing amount of attention has been given by investigators of animal behaviour to patterns of birds song–their control, development, and evolutionary significance. What about the development of song in birds? There appear to be a wide diversity of developmental strategies leading to the diversity of songs and calls with which we are all familiar. From comprehensive analyses of a number of bird species there appear to be at least two major strategies for song development; (1) imitation of the songs of others, particularly of adult conspecifics; and (2) invention or improvisation. Underlying both strategies are the issues of what type of templates exist, which provide some genetic basis for and help to shape or guide the song learning process, and the possible existence of a sensitive phase for development of the song repertoire. Several examples should help explain song development.

**In Marsh Wern**

The marsh wren (*Cistothorus plaustris*) has been used extensively in studies of song learning. Male marsh wrens in nature sing over one hundred types of songs; neighbouring males generally sing identical song types; and males often interact by countersinging with one another using the same song type. To test development of song learning males were reared in special housing conditions where the songs they heard could be completely controlled. Males were played specially prepared tutor tapes of songs; each tape contained nine different songs. Males were exposed to one set of tapes for the period 15 to 65 days of age, a second tape for age 65 to 115 days, and a third tape the following spring. The males learned, by imitation, the nine songs on the tapes to which they were exposed prior to 65 days of age, but did not learn the songs on the subsequent two tapes. The males never sang any invented songs. Further investigation revealed a more refined estimate for the peak sensitive period, which falls between about days 35 to 55 of age. In addition, males learned song types played for either three days or nine days. In these additional tests some males improvised some songs, presumably from elements of the songs on the tutor tapes. The learning situation used for these marsh wrens involved only tutor tapes played over loudspeakers.

In another test young wrens were first exposed to a number of song type vis tutor tapes and then were given a period of social interaction with adult males with varied song repertoires. The period of social interaction occurred in the fall of the first year for some birds and not until the following spring for others. Data on song repertories for these birds indicate that song learning can occur early from the tapes, or at either of the period of exposure to adult males. Hence, there is some flexibility in terms of song learning. These findings also make it clear that we should be careful in using only rather artificial stimuli like loudspeakers. Social interaction has also been shown to be a critical factor in language acquisition in human children.

**In Male Swamp Sparrow**

A longitudinal study of song development has been carried out for male swamp sparrows (*Melospiza georgiana*) that were hatched in the wild and brought into the laboratory. Beginning at 16 to 26 days of age the birds were given song training twice per day for forty days with songs typical for the species. Recordings were made of the songs of all birds once each week, commencing shortly after training, when birds were just over three months old, and continued until they were over one year of age. When these recordings were analysed, a seven-stage sequence of song development was discernible. The syllables used in training are shown at the top of the figure. Young birds began by singing what *Marler* and *Peters* call subsong at an average age of 272 days. They progressed through sub-plastic song and began to sing the plastic song at an average of 299 days age. Crystallised song began, on average, at 334 days of age. During the course of this developmental sequence, the duration of the song decreased. Syllabic structures began to emerge during sub-plastic song. Analysis of the syllables revealed that about 30 per cent were imitations of the training songs and 70 per cent were inventions or improvisations.

By the time of emergence of the crystallised song the number of syllables sung was only 23 per cent of the potential repertoire; both imitated and improvised syllables were included in the crystallised song of most birds. During development it was not unusual to record songs characteristic of more than one stage of the sequence on the same day. However, once the singing of crystallised song began, the birds rarely reverted to an earlier stage. *Marler* and *Peters* (1982) hypothesize that the pattern of song development noted for the swamp sparrow may be characteristic of many song bird species.

**In Female Mice**

Puberty is a critical event in the lives of most organisms–for many, sexual maturation marks the onset of reproductive behaviour and the production of progeny. Puberty is also important in terms of the population biology of many species. For house mice (*Mus musculus*) the deme structure generally involves one to several adult males, three to seven females, and their pups and juveniles. Some juvenile females may disperse from the natal site, whereas other may remain within the deme, but virtually all juvenile males disperse. Within this social context the timing of puberty in female mice has been shown to be significantly affected by various social and environmental factors.

Puberty is measured by the occurrence of first estrus; the heightened levels of estrogen result in cornification of the cells that line the vagina. The phenomenon can be detected by microscopic examination of a vaginal smear. The presence of a mature male mouse or daily exposure to urine from mature males accelerates the onset of puberty in young female mice, as does daily exposure to urine from lactating females and urine from single-caged females, regardless of age, results in delays in the onset of puberty. Mice exposed daily to clean bedding or to urine from single caged diestrus females reach puberty at ages that are intermediate between the acceleration and delay effects produced by exposure to urine from the other sources.

**Table 2.2. Mean Ages and Ranges for Sexual Maturation of female House Mice (Mus musculus) Given Various Treatments with Urinary Chemosignals**

| *Treatment* | *Mean Age at Puberty (days)* | *Age Range (days)* |
|---|---|---|
| Control females–exposed to water treatment daily | 35 | 29-42 |
| Females caged with an adult male | 27 | 24-31 |
| Females exposed daily to urine from adult males | 31 | 27-36 |
| Females exposed daily to urine from grouped females | 41 | 36-45 |
| Females exposed daily to urine from estrous females | 30 | 26-36 |
| Females exposed daily to urine from di-estrous females | 36 | 30-41 |
| Females exposed daily to urine from lactating females | 30 | 27-35 |
| Females exposed daily to urine from pregnant females | 31 | 26-36 |

These effects have been demonstrated in both laboratory and wild stocks of house mice and several of the effects have been replicated in stocks of wild *Mus musculus* maintained in the cloverleaf islands of superhighways. When mice are exposed simultaneously to urine from two or three sources the outcome depends on which sources were used. For example, if treatment involves any exposure to urine from grouped females, puberty is delayed in young test females–regardless of what other sources are used. If all donor sources involve urine which accelerate puberty, then the combination treatments result in earlier maturation, but not any earlier than using one of the sources alone. Diet and daylight have also been shown to affect the timing of puberty in female house mice.

Hence, a variety of contextual cues from the developing mouse's environment can influence internal physiological events. These effects, in turn, have important consequences for the onset of reproductive behaviour in the mice. The chemosignal effects will also influence the length of the interval between a mouse's birth and when it starts to reproduce–called *generation time.* Generation time is a key element in determining the rate of growth or decline in the numbers of young mice entering the population over given length of time.

## In Insects

### *Fruit Flies*

Fruit flies (*Drosophila*) have been used in a wide variety of investigations of the development of behaviour; studies have been done both on larvae and on adults utilising field and laboratory conditions. The primary activity of larvae is feeding. As the larva moves across the food surface it provides with its mouthparts, ingesting food with each cycle of extension and retraction of its body. The larva of *D. malanogaster* as adult flies. The rate of feeding reaches a peak early in the third instar and then declines during that instar. Before pupation the larva may migrate to a pupation site away from the food source, or it may remain at the food source during pupation, the pattern varies for different species and sometimes even within a species.

Some species may burrow into the soil to pupate, possibly to decrease the risk of predation. The adults of many *Drosophila* species, particularly females, have been studied more extensively than larvae–adults perform a considerably greater number of behaviour patterns. Two major behaviours related to reproduction undergo developmental changes in females, sexual receptivity and oviposition. *Manning* studied sexual receptivity in *D. melanogaster*. As the corpora allata, which release juvenile hormone (JH), and ovaries grow larger, the female becomes receptive. This generally occurs about 48 hours after the *eclosion* of the adult fly from the pupa.

Immature females reject courting males that attempt copulation. One hypothesis relating the physiological and behavioural events postulates that JH may affect the brain directly or indirectly, lowering the threshold for sexual receptivity. If activity functioning corpora allata are implanted in females at the pupal stage, the emergent flies exhibit earlier sexual receptivity and have larger ovaries than nonimplanted controls. After mating, females become unreceptive again and soon the oviposition behaviours increase. The turning off of receptivity appears due to at least three factors: the presence of sperm in the female' receptacles, the act of copulation itself, and a secretion from the paragonial gland of males. The increase in rates of oviposition is probable due to the increased size of the ovaries, which provide information to the nervous system

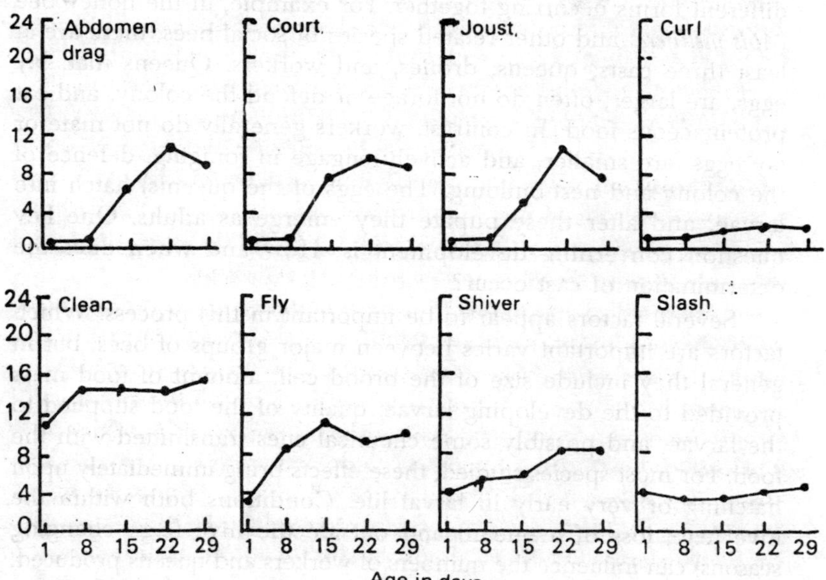


*Fig. 2.12. Frequencies (X̄ ± -SE) of eight behaviours in Drosophila grimshawi males at ages of 1, 8, 15, 22 and 29 days, post-eclosion.*

via stretch receptors, and to the presence of male paragonial gland secretion. *Ringo* discusses the development and maturation processes in females and males of a group of *Drosophila* species inhabiting the *Hawaiian islands.* Development takes a long time in these species, lasting for days or weeks. For males of *D. grimshawi* a series of eight behaviour patterns were observed and recorded at four ages during a one-month period following eclosion.

In general, the diversity of behaviours observed increased with age, and the relative frequency of each behaviour increased with age. Males of this species from leks, in which groups of males display communally and mate with females attracted to the lek. For days 15 and 22 there were pronounced increases in sexual and agonistic behaviour–courting, jousting, and abdomen drag. The increases in behaviour correspond to the time when the males are most likely to be competing for copulations.

***Honey bees***

Many bee species live in large colonies in which there are castes–physiologically, behaviourally, and often morphologically different forms occurring together. For example, in the honey bee (*Apis mellifera*) and other related species of social bees, there are at least three casts; queens, drones, and workers. Queens mat, lay eggs, are larger, often do not forage or defend the colony, and eat proteinaceous food. In contrast, workers generally do not mate or lay eggs, are smaller, and actively engage in foraging, defence of the colony and nest-building. The eggs of the queen(s) hatch into larvae, and after these pupate they emerge as adults. One key question concerning development is: How and when does the determination of cast occur?

Several factors appear to be important in this process. Which factors are important varies between major groups of bees, but in general they include size of the brood cell, amount of food mass provided to the developing larvae, quality of the food supplied to the larvae, and possibly some chemical cues transmitted with the food. For most species studied, these effects bring immediately upon hatching or very early in larval life. Conditions both within the hive (e.g., loss of a queen) and outside the hive (e.g., changing seasons) can influence the numbers of workers and queens produced. Most worker bees of there social species undergo changes in behaviour as they develop which correspond to changes in their functional roles within the colony or hive. Resting and patrolling occur throughout the twenty-four days surveyed.

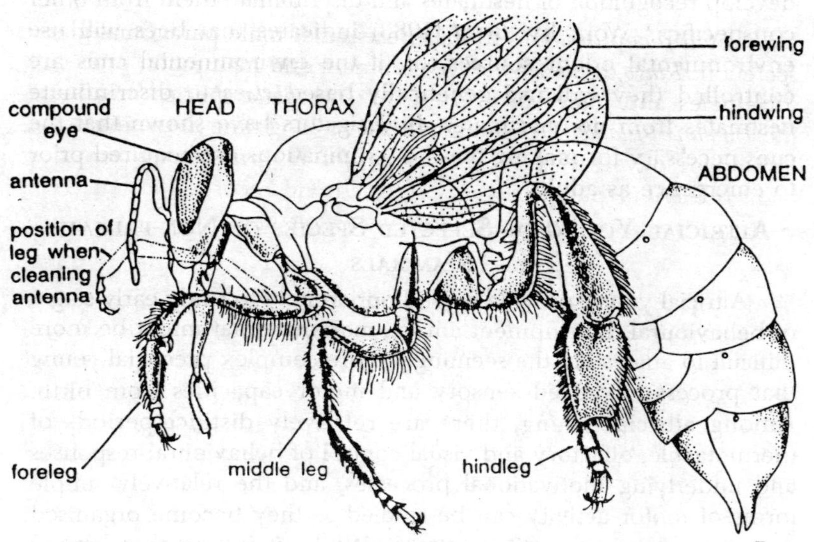


*Fig. 2.13. Worker honey bee (Lateral view).*

Many other activities occur in a sequential pattern, beginning with high levels of cell clearing in the early days. This is followed by activities related to the comb and tending the brood. For the last portion of their lifespan the workers became foragers; the average worker bee living in a temperate zone climate survives to the age of six weeks. For many of the activities there is some overlap–bees shift among behaviours off all types on a given day, up to the start of foraging activity. Several physiological changes are correlated with the shifting patterns of functional roles. Young bees have enlarged hypopharyngeal glands. These produce a major component of the bee milk, part of the diet fed to larvae. Levels of secretion for invertase, an enzyme involved in the conversion of nectar to honey, are highest during the middle portion of the lifespan.

Wax glands are small in the youngest workers, but this is followed by a gradual increase to maximum size by about days 16 to 18, and then the wax glands decline rapidly. These changes correspond with the higher levels of comb-related activities during this age range. Many other traits follow similar patterns of correlation between developmental events throughout the life of the bee and its changing functional activities in the hive. Most social bees actively guard and defend their colonies. How do they develop recognition of nestmates and discriminate them from other conspecifics? Work by *Breed* (1983) indicates that bees will use environmental odour sources, or, if the environmental cues are controlled they will use genetically based cues to discriminate nestmates from non-nestmates. Investigators have shown that the cues necessary for making these discriminations are acquired prior to emergence as adults.

## Altricial Young of Selected Species of Non-primate Mammals

Altricial young provide a good opportunity to study early stages of behavioural development among mammals that might be more difficult to analyse in the seemingly more complex precocial young that process advanced sensory and motor capacities from birth. Among altricial young, there are relatively distinct periods of thermotactile, olfactory and visual control of behavioural responses and underlying motivational processes, and the relatively simple forms of motor activity can be studied as they become organised into more complex action patterns While, it is true that limited motor abilities may hide advanced sensory and integrative processes in altricial young, it is equally true that advanced sensory and motor abilities may mask simple behavioural capacities in percocial young.

The problem is, therefore, to determine, through analytical experiment procedures and theoretical synthesis of the findings, the natural of behaviour organisation at each stage of development. We are not prepared to develop detailed models of the behavioural organisation of young at the various stages but we can sketch the major lines alone which development proceeds. Among altricial young these lines are clear from many good descriptions of the development of behaviour toward the mother, siblings and nest or home site in a variety of mammalian species. Suckling and nom-suckling contact with the mother, and huddling with siblings in the nest, take an early lead in behavioural development among altricial young.

At first the mother takes the initiative in suckling, but gradually the young contribute increasingly to the initiation and termination of feeding sessions, until the relative contribution of each shifts and the young play the dominant role in the feeding interaction until weaning intervenes. Similarly, huddling among siblings is a relatively simple behaviour pattern while the young are about more freely, behavioural interactions among siblings become more complex: young become capable of many more responses to one another and respond increasingly to individuals and their characteristics rather than to group characteristics. Initially, the young are either passively confined to the home area or are actively oriented to it but as development proceeds they become able to move more freely in relation to it at a distance. This early form of orientation to the home area wanes after a period and the young's relationship to its socially conditioned environment is based upon its relationship to its companions and to various specific functions that emerge as development proceeds.

**Suckling**

Among all species of altricial mammals suckling is initiated shortly after birth. The newborn actually begin to nuzzle the mother's nipple region during parturition, when given the opportunity in the intervals between births. Kittens, however, are only rarely successful in attaching to nipples and this is very likely the case in other altricial species. *Ewer's* suggestion that the delay of an hour or so *post partum* in the onset of suckling in kittens is based upon the absence of a special releasing stimulus for nipple grasping seems unnecessarily elaborate.

Kittens have little opportunity to attach to nipples during parturition even when the mother is not actively giving birth. The mouth is engaged in a variety of activities such as licking the kittens, eating the placentae, licking birth fluids from her fur and the delivery site, etc., which make it difficult for kittens to attach to nipples. Not until parturition is completed does the female make herself available for feeding. The position assumed after parturition by the exhausted mother among cats, dogs and other small mammals facilitates nipple searching and grasping by the newborn, her limbs outstretched forming an oval-shaped corral in which the young are confined. Comparable positions are assumed by rat, hamster, and rabbit mothers. Most successful artificial mothers have been modelled after the immediately post-partum nursing female.

In this position the mother presents a variety of attractive thermal, tactile and olfactory stimuli. The effects of these stimuli on early suckling have been investigated through the presentation of one or several of the compónent stimuli under controlled conditions in artificial brooders. Among the important features of the mother's body that stimulate approach and nuzzling appear to be the fur-textured surface, and the rounded shape with few entrapping crevices or projections that distract newborn from locating the nipples and surrounding areola. Body surface temperature in the range of 32 to 39°C, as measured in various species, and as provided in artificial mothers, has proved attractive to newborn rats, rabbits, puppies and kittens.

The moist surface of a pulsating plastic tube was most attractive to brooder-reared rat pups in the study by Thoman & Arnold. In addition to being attracted to the mother's nipple region by these stimuli; newborn kittens are led along certain paths in their nuzzling of her body by patterns of thermal and tactile stimuli. *N. Freeman* found a broad but well-defined thermal gradient extending over the entire body surface of the lactating cat that could provide a basis for locating the nipples in newborn kittens. When on a floor surface at 30.2°C a thermal gradient extended from the mother's limbs, which had a surface temperature of 33.3°C, to her nipples which were at 37°C. The pattern of fur growth offers additional directional stimulation : when the mother is lying on her side, crawling against the grain leads kittens upward from the midline to the lateral nipples. Moreover, the areola surrounding the nipples, with its bare, warm moist surface, attracts kittens.

It has been suggested that the pattern of hair growth on the belly of the pig provides paths which piglets follow in finding nipples. Static features of the mother and more active ones, such as her licking that steers kittens to her ventrum, offer attractive stimuli which initially bring newborn into contact with her. More detailed and patterned features lead to them to the nipples.

## Suckling and Thermotactile Stimuli

Thermal and tactile stimuli provide the basis for the earliest suckling approaches to the mother in kittens, rat pups, and newly born rabbits and puppies. In the kitten, the snout and lips are supplied with low threshold thermotactile receptors and greater anterior-end sensitivity to such stimuli appears to be the general rule among newly born altricial young. The snout functions as a probe as it moved by side-to-side head movements during crawling, a characteristic of kittens and other altricial young that move in this manner. These movements enable the newborn effectively to scan a fan-shaped region in front of it.

Contact with a low-intensity tactile or thermal stimulus slows the return head swing and the newborn veers in the direction of the stimulus by repeated movements of the contralateral forelimb and cessation of ipsilateral forelimb movements. On the other hand, contact with a strong tactile stimulus (e.g., wire texture floor surface, or excessively warm or cool stimulus (e.g., overheated or cooled floor surface or overly warm or cool littermate, initiates head withdrawal followed by cessation of forward movement and veering away.

If the stimulus persists, the newborn pivots a half circle and crawls away from it. This withdrawal response, which can also be elicited by a puff of air directed at the newborn's face, has been used to establish early escape and avoidance learning in newborn of several altricial species. At this early stage, thermal and tactile stimuli in the litter situation also stimulate newborns to activity which eventuates in their crawling to the mother. *Okon* has shown the distress-inducing effects of strong tactile stimulation and extreme thermal conditions upon hamster, rat, and mouse pups, as measured by a high rate of audible and ultrasonic calling. This builds up during the first weeks after birth then declines at the end of the second and beginning of the third week. Although rough handling or tail pinching are most effective in eliciting ultrasonic calling, it is particularly significant that loss of contact with tactile stimulation (i.e., even nesting material) stimulates calling in three days old rat pups and young hamsters.

Similarly, cooling caused by exposing three to nine day old pups to a 22°C ambient temperature stimulates a significant rise in the proportion of animals that vocalise and in the rate of ultrasonic vocalisation. Newborn rabbits are stimulated to activity at 20-25°C ambient temperatures which cause a decline in body temperature of 9°C.

Among puppies and kittens loss of contact with tactile stimulation from the mother and littermates stimulates an increase in activity that usually consists of pivoting and calling. In puppies, quiescence is restored when head and body contact is made with a soft tactile stimulus, and a most effective stimulus for quieting kittens that have lost contact with the mother and litter is the weak tactile stimulation of the face and forehead provided by a canopy of soft bunting. Cooling is as effective as tactile stimulation in a arousing puppies and kittens to activity and in stimulating calling.

In an ambient temperature of 30°C puppies remain quiescent, breathing lightly in a relaxed sleeping posture, but if the *ambient* temperature falls to 27°C or rises to 32.5°C, they become alert and restless, being to call, and exhibit rapid breathing. Kittens placed on a cool surface (18°C) become active and call during the first week. Many crawl to a warmer area if this is available, but others become so active they simply walk off the table surface. Newborns that lose contact with the mother therefore become active as a result of both the loss of tactile stimulation and exposure to a lowered ambient temperature. They begin to pivot and circle, emitting calls to which the mother often responds by bringing her body closer and into contact with the young or by licking them.

The newborn is likely to make contact with the floor surface around the mother which is warmed by her body and with her extended fore- and hind-limbs as well as the warm furry surface of her nipple region. These ensure that the newborn will regain contact and warmth and if it has not fed for a time, that it will initiate nipple-locating movements and finally suckle from the mother.

**Nipple Grasping and Sucking : Specialised Responses**

Nipple grasping and sucking are specialised responses of the newborn to features of the nipples and the surrounding areola. The specific nipple characteristics which enable newborns to grasp them have not yet been established for any altricial species and so newborns do not initially attach spontaneously to the nipples of artificial brooders. The projection of the nipple from the surrounding mound of bare skin (areola) is an important feature enabling kittens to locate and grasp the nipples. While nuzzling the mother's fur they crawl upward on her body but upon reaching the areola region their nuzzling abruptly changes to gentle nose tapping and forward crawling stops. As they contact the projecting nipple with the nose and lips, the head is withdrawn and raised and the mouth is opened; the nipple is grasped by a forward head lunge with mouth open, and often several such lunges are made before the nipple is centered in the kitten's mouth. This is a variable response even in newborn kittens.

Brooder reared kittens develop a different pattern in adapting to the longer and more flexible rubber nipple used. They approach in a similar way but upon making contact they position the nipple at one side of the mouth, move the head sideways, thus bending the nipple, then allow it to spring back into the mouth. If brooder nipples are placed in rounded depressions on the brooder surface, kittens find it difficult to attach to them. The reason appears to be that when probing the depression, the kittens are stimulated more strongly around the face by the edges of the depression than on the nose and mouth by the nipple. Face stimulation stimulates crawling and strong pushing into the depression but appears to inhibit a response to the nipple.

The location of the nipple on a mound, as in the mother, therefore has the opposite effects: crawling is inhibited by the nose stimulation and nipple grasping is elicited. Each kitten rapidly develops a preference for either a single nipple or a pair of nipples. Texture differences on the surface surrounding the nipples may play a role in this discrimination., since R.J. Woll has shown, in brooder-reared kittens, that a discrimination between differently textured nipples flanges, one providing milk and the other without milk, can be learned by the second or third day. In rat pups tactile stimulation of the upper lip appears to be the stimulus for nipple grasping. Pups, whose upper lips had been desensitised, failed to grasp the nipples although they nuzzled them when placed on the nipple region of anaesthetised mothers. When they were placed on nipples, however, they sucked normally.

Once the nipple is grasped and sucking brings, young are extremely sensitive to tactile and perhaps thermal stimuli within the oral cavity. A series of postural changes during suckling has been described in rat pups, in which changing nipple stimulation and finally milk ejection stimulate first treading then stretching responses. Treading anticipates actual milk ejection and may be in response to subtle oral tactile or thermal stimuli resulting from milk let-down within the mammary gland. In human mothers, oxytocin released in response to infant crying raises breast temperature by 1°C as a result of milk let-down. The likelihood of thermal stimulation playing a role in nipple grasping arises from *N. Freeman's* finding in the lactating cat that nipple temperature is highly (37°C) than the nearby surrounding skin (34.7°C) and G.H. Rose's observation that warming the nipple and flange of an artificial mother aids the newly born kitten to find and grasp the nipple. Stanley and his associates have shown that the newborn puppy is able to modify its sucking in accordance with the rate and pattern of milk flow, probably on the basis of oral tactile stimulation.

## Suckling and Intraorganic Stimulation: Early Stage of Development

The initial and terminal of suckling cannot be explained entirely by reference to exteroceptive thermal and tactile stimuli from the mother even in the early stage of development. Young of all altricial species exhibit a periodicity of feeding that appears to depend on internal sources of stimulation generally labelled, in aggregate, as hunger motivation. The most obvious during suckling is the milk which fills the newborn's stomach. It has been assumed that filling of the stomach terminates suckling as a result of stimuli arising from muscular distention of sensory receptors in the stomach, and, conversely, that an empty stomach stimulates the young to initiate suckling and to be responsive to nursing stimuli. *James* was the first to show that in young puppies (i.e., less than 26 days) suckling was not inhibited by preloading their stomachs with adequate amounts of milk. Latencies to initiate suckling and the duration of suckling were not different in preloaded and non-preloaded puppies which had been separated from their mothers for up to 3 hours, although preloaded puppies consumed less milk.

Further study showed that preloading to the extent of overloading (i.e., milk regurgitated through the mouth) could inhibit suckling and reduce milk intake, but strong effects had to be made with these fully preloaded puppies to keep them awake. Even though they attached to nipples, once they were released after being hand held and stimulated they released the nipples and fell asleep. Thus, the stomach probably does not directly regulate the responsiveness of young puppies to the exteroceptive nursing stimuli of the mother. *Hall* has recently shown this even more clearly with rat pups during the first 10 days, using the anaesthetized mother to test for sucking.

*Hall* deprived one group of pups of suckling for 22 hours and removed a second group from the mother just before testing and anaesthetisation. Both deprived and non-deprived pups were placed on the mother's exposed nipple region and their readiness to attach and suckle was measured. There level of general activity was also measured. During the first 10 days, deprived and non-deprived pups attached with the same latencies. Latencies for attaching decline in both groups from just below 200 seconds to between 50 and 75 seconds; activity levels did not differ substantially between the two groups for most of the 10 days period.

This study shows that readiness to suckle was not dependent upon any differences in stomach loading that resulted from differences in the recency of suckling. Anaesthetised rat mothers do not release milk in response to sucking because of the inhibition of oxytocin release necessary for milk let-down. In Hall's study, therefore, pups sucked from dry nipples yet they continued to suck for the remainder of the five minutes test period. In this early period, therefore, ingestion of milk is not an essential stimulus for sucking; Hall reports that pups eight days of age and younger suck for at least six hours, continuously, despite the fact that they obtain no milk throughout this period. Similar findings have been described for kittens during the first three weeks after birth. Two groups of kittens were studied : one was presented with a normal lactating mother and the other with an anaesthetised non-lactating female in whom sucking did not, of course, cause milk release.

Both groups were fed adequate amounts of milk by tube directly into the stomach at four hour intervals and were exposed to the 'mothers' for six hours each day, the exposure starting two hours before a scheduled tube-feeding. During the first three weeks the dry-sucking group sucked as frequently and for as long as the group that obtained milk during sucking. Both groups sucked (i.e., were attached to nipples) for between 31 and 42 minutes of the first hour, indicating that latencies to attach were rather short and did not differ between groups.

At the end of two hours, both groups were tube fed according to schedule and the effect was to reduce sucking during the third hour, equally in both groups, but not eliminate it. This study suggests that milk intake is not essential either for the initiation of suckling during the first three weeks or for maintaining sucking. We shall see that in puppies, rat pups, and kittens, the situation changes as the young become older and milk intake plays an important role in sucking. Yet there is a periodicity in feeding even during the early period that is not based entirely upon the mother's behaviour since it appears in brooder-reared kittens as well. Among kittens and rat pups the young sleep during sucking though they may remain attached to the nipples, and when the mother raises to leave, some are so strongly attached that they are dragged from the nest. The ingestion of milk plays a role in sucking but its role is indirect. Following stomach filling with milk the young may be lulled to sleep, as a result of a sharp reduction in general arousal perhaps through inhibitory effects of stimuli from the distended stomach.

Sucking may be terminated, therefore, or fail to be initiated or only weakly initiated, in young preloaded with excessive amounts of milk. During the first 10 to 15 minutes after preloading with milk, kittens exhibited an overall lethargy that made them unresponsive to all forms of stimulation and preclude any response to a nursing mother. There was then gradual recovery and the kittens became more responsive to stimuli and often began to initiate suckling approaches with nipple grasping and sucking. Suckling has many intraorganic effects some of which derive from milk ingestion and others from exteroceptive stimuli received during contact with the mother.

*Hofer* and his associates have uncovered an effect of milk intake on cardiac function. Two week old rat pups, separated from their mother overnight, show a decline in heart rate that cannot be attributed to a decline in body temperature. These investigators have shown that only tube feeding at regular intervals (i.e., one or four hour intervals) prevents this decline : *Koch & Arnold* have extended these findings to pups four to 14 days of age. In addition they have found in four to 10 day old pups that even non-suckling contact with the mother by tube-fed pups accelerates cardiac functioning more that tube feeding alone. Since rat pups in this age range are unable to thermoregulate, the effect may be based upon the thermal insulation provided by contact with the non-suckling female as well as contact with a source of warmth well above the pups' body temperature. Rabbit young suffer a loss in body temperature when isolated in 20-25°C ambient temperature and contact with the mother produces an immediate rise in the young's body temperature. This suggests that suckling exerts an important influence on the body temperature of young partly through body contact and partly through the warming effects of milk: full body contact with the mother has similar or even greater effects. It is perhaps significant that others from a different perspective have proposed that maternal behaviour among mammals has evolve primarily in relation to the thermoregulatory needs of altricial young and that nursing is a secondary development which evolved later.

Nevertheless, a number of investigators have shown that altricial newborn of a number of species can learn discrimination of various sorts on the basis of milk reinforcement of sucking. Tactile, taste, and olfactory discriminations have been established and sucking has been 'shaped' by the presentation of milk. If the effects of reinforcement are related to the underlying motivational conditions, then in these young, reinforcement must act in a different way than in adult animals with well-defined motivational systems. The nature of this difference, however, is not yet known.

**Suckling and Olfaction**

Olfaction enters early into the suckling pattern of altricial newborns but it may be several days before it plays a significant role in the approach to the mother and in nipple searching. As early as three days of age, rat pups show evidence of responding to the odour of a lactating female or of nest deposits, and on the fifty day pups show behavioural arrest and EEG synchronisation in response to material odour. By the ninth day the evidence of response to maternal odour is clear. Bilateral bulbectomy on the second day has a severe and irreversible effect on suckling in rat pups, resulting in a gradual weight loss beginning one or two days after surgery and 80% mortality by the sixth day.

Pup approaches to the mother when she enters the nest and interactions with littermates are only mildly affected by bulbectomy. At later ages also, bulbectomy produces a severe reduction in suckling and consequent weight loss, with high mortality but mild effects on social interactions with the mother and the littermates. The effect of bulbectomy appears to be specific to nipple grasping since approaches to the mother and nuzzling in her fur are not different in bulbectomized and normal or sham-operated control pups. This suggests an early role for olfaction in nipple grasping. *Kovach* and *Kling* similarly found that bulbectomized kittens could not suckle with the mother but were able to suckle when placed on an artificial nipple. The rabbit pup, kitten and puppy show evidence of use of olfaction in their responses to the mother during the first week. Artificial odours have been applied to the mother's nipple region and later tested alone to determine whether, in the course of suckling, young rabbit pups and puppies are influenced in their behaviour by the presence of the odour.

Rabbit young thus exposed to odours during the first five days of suckling began to respond positively to the odour alone on the second or third day and puppies exposed similarly responded immediately when tested for the first time on the sixth day. Among kittens any of the three functional pairs of nipples may be suckled by each of the kittens of a litter during the first day *post partum.* By the second and third day, however, each kitten confines its suckling to either a single nipple or a pair of nipples. Once nipple position preference are established nose contact with non-preferred nipples does not elicit nipple grasping as it did on the first day.

To determine whether olfactory cues could be the basis for these preferences, brooder-reared kittens were presented with two nipples and flanges, one of which provided milk while the other was blind. Two-day old kittens were tested for their ability to establish as olfactory discrimination. Within a day or two the kittens were able to discriminate between two artificial odours: they approached and suckled almost exclusively the nipple-flange combination with the positive odour. The ability of rat pups to approach the mother on the basis of an olfactory stimulus has been clearly established by *Moltz and Leon.* From 14 days pups respond to the odour of lactating mothers in preference to that of non-lactating females.

The odour emanates from a volatile component of the mother's faces which the young ingest, and if mothers are fed different diets the offspring can distinguish the odour of their own mother. Under these conditions, the odour of a strange mother has no special attraction for them, including that material odour acquires its attractiveness through experience. Olfactory stimuli appear gradually to become a component of the young's approach and suckling response to the mother in these altricial species. From the beginning, olfactory stimuli are able to influence the newborn's behaviour. Ammonia fumes (which may in fact have irritating tactile effects) stimulate pivoting and vigorous crawling in newborn rats.

There are, undoubtedly, odours to which newborn respond positively from birth onward, particularly those related to birth fluids etc, but, in the main, odour appears to acquire their significance through association with thermotactile stimuli that accompany them. When the development of olfactory responses to the mother's body and particularly to those regions involved in nursing, certain changes occur in the suckling pattern.

1. Suckling becomes more specifically a response to the newborn's own mother and to specific aspects of the mother's nipple region (e.g., development of nipple position preference).
2. Since olfactory stimuli from the mother may reach the young before actual contact is made with her, the young may initiate the suckling approach while the mother is still at a short distance and thus they may anticipate her approach.

3. Since olfactory stimuli may be spread outside the nest or home site, familiarity with maternal odour may enable young to widen the range of their activity to these areas.
4. Since olfactory stimuli may also be deposited on littermates, young may also begin to respond to their siblings on this basis.

## Suckling and Vision

At the time vision becomes functional in altricial young, the suckling pattern is well established on the basis of olfactory and thermotactile stimuli. These stimuli, however, can only be received by young when the mother is close by or in actual contact with them. Use of vision enables young to perceive the mother at a distance and approach her for suckling. The onset of visually stimulated suckling behaviour is usually taken as the beginning .of distant approaches to the mother.

In the rat this begins between the fourteenth and sixteenth days, in the hamster around the second week, in the puppy around the fifteenth day, in the rabbit around the tenth to fourteenth days, and in the kitten around the seventeenth day. It must be kept in mind, however, that vision may begin to function earlier in suckling but at short distances where it would be difficult without specific examination to distinguish between a thermotactile, olfactory, or visual basis for the young's approach. While there is considerable descriptive evidence of visually guided approaches to the mother beginning sometime after eye-opening in the species we have been discussing, experimental evidence if often lacking.

In the kitten eye-opening occurs around the seventh to ninth day and there is evidence of the onset of visual functioning between the fourteenth and seventeenth days. Visually guided approaches to the mother from a distance start around the seventeenth day and are fully established by the twenty-first. In the 10 day old rabbit pup an odour associated with the mother initiates approach to the mother which is then guided by visual stimulation. The implication which is then guided by visual stimulation. The implication of the studies on maternal pheromonal stimulation in the rat by *Molta and Leon* is that, from around the fourteenth day, pups identify a lactating female by odour but their approach to her is guided by visual stimulation.

*Altman* and his colleagues have shown that rats orient to the mother and littermates in the home cage, presumably on the basis of visual supplement by olfactory stimuli, during the third week. The beginning of visual functioning in the rat pup has been found on the fourteenth day by *Turner* who studied pups' ability to find an exit door, marked by a vertical stripe, in a circular field.

Pups improved rapidly from the fourteenth day onward in their ability to go from the centre of the field directly toward the door, a distance of 13 inches (33 cm). Visually guided approaches to the mother in puppies described by *Rhein-gold* as starting around the fifteenth day and increasing therefore, are correlated with visually evoked responses in the cortex which assume the adult form at around the same age. By four weeks of age, puppies will approach a two-dimensional drawing of a female, paying particular attention to the head and flank, and at a slightly older age to the mammary region, when they are hungry. The onset of visually stimulated approaches to the mother marks an important stage in the young's development with implications for the organisation of suckling behaviour.

It is no doubt significant that the ages at which young increasingly initiate suckling approaches to the mother from a distance in the rat pup, puppy, and kitten correspond roughly with the ages at which intraorganic stimuli becomes increasingly important in the regulation of feeding. If the young approaches the mother when she is unresponsive or not easily available it is likely to result in either failure to suckle or in rejection by the mother. It is during this stage that maternal rejection or evasion of the young become prominent. Sucking approaches made when the mother is receptive result, on the other hand, in her co-operating with them by assuming a nursing position. Young gradually become sensitive to subtle features of the mother's behaviour indicating her readiness or unwillingness to nurse and adjust their suckling approaches accordingly. Moreover, each young approaches the mother to suckle individually, whereas earlier the mother's approach had stimulated the entire litter simultaneously and all young suckled together. Once nursing is initiated with one offspring other young often join in, but each young acts on the basis of its own perception of the situation rather than to the stimuli that initiated suckling in its sibling.

## Suckling and Intraorganic Stimulation: Late Stage

Between the tenth and fourteenth day an important change takes place in the intraorganic stimulation of suckling among rat pups.

Around this age the latency to initiate suckling (with an anaesthetised mother) begins to differ between pups that have been deprived of suckling for 22 hours and those that have only recently been removed from their mothers. Latencies are very short for the deprived pups but range above 200 seconds for recently fed pups. Moreover, failure to receive milk from the anaesthetised mother, which had little effect on maintenance of suckling earlier, now begins to affect suckling: pups shift from one nipple to another, within the five minutes suckling test, presumably in response to the dry nipple. Similarly, around the third week of age, kittens suckling from a dry mother initiate nuzzling of the mother's nipples with the same frequency as kittens suckling from a lactating mother, but the frequency of attaching and suckling declines rapidly from the earlier period and the duration of sucking also declines to 20% of the previous period. Meanwhile kittens suckling from a lactating mother continues to suckle for extended periods.

After 30 days of age, preloading the stomach of puppies with a milk-chow mixture has the same effect of reducing or eliminating further eating from a dish as the same amount of food consumed voluntarily by control puppies. This contrasts with the earlier finding that preloading the stomach with either milk alone or a similar mixture has no effect on suckling. Indications are, therefore, that arousal produced by intraorganic stimulation has become more specific than earlier so that exteroceptive stimulation received before or during suckling is not sufficient to initiate or maintain suckling. Milk intake and its intraorganic effects have become an essential part of the sucking pattern.

At this stage of development, stomach filling or its absence has become a more focal source of arousal than earlier and it determines to a large extent the young's response to the exteroceptive stimuli associated with suckling. The kind of exteroceptive stimuli that aroused suckling approaches undergo a change during this period. Among rat pups this is the period during which suckling approaches are increasingly initiated by the pups from a distance, in response to olfactory and visual stimuli from the mother. Among kittens there is the beginning of kitten-initiated suckling approaches to the mother, starting around three weeks of age.

At this stage suckling approaches are increasingly determined by the past experience which the young has had with the mother and less by immediate stimuli for suckling. This was shown in a study by *Kovach and Kling* in which they found the kittens reared in isolation from an early age until about three weeks and fed by stomach tube to prevent suckling, were unable to initiate suckling when returned to their mothers. While they attributed this to the waning of the suckling reflex, our analysis indicates that at three weeks off age the sucking reflex is only a small part of the suckling pattern.

Moreover, as part of an organised suckling pattern it initiated mainly on the basis of intraorganic stimuli rather than the previously effective proximal exteroceptive stimuli. In a study comparable to that of *Kovach and Kling* kittens were reared in isolation from one to three weeks of age with feeding by suckling from an artificial rubber nipple mounted on a brooder. When these kittens, in whom suckling had maintained as an active component of the feeding pattern, were returned to their mothers there was a delay of about 20 hours in initiating sucking. The previously isolated kittens had no difficulty in establishing sustained contact with the mother, which in all cases took less than three hours and in most kittens was accomplished within the first hour. Their difficulty lay in perceiving the mother as an object to be suckled. They crawled over her body, slept in close contact with her nipple region and often nuzzled, but not until they had been without milk for many hours did they finally exhibit nipple grasping and suckling.

In these brooder-reared kittens the suckling pattern had been organised in relation to the brooder and artificial nipple, which differed in many ways from the mother. Under comparable conditions in the brooder situation, these kittens would initiate suckling within seconds after being returned to it. As young develop, therefore, the relative influence of exteroceptive and intraorganic factors in the regulation of suckling shifts in favour of intraorganic factors. Moreover arousal undergoes changes: It can no longer be described simply in terms of general arousal. Arousal has become more specific as a result of experience in the litter situation and of neutral maturation which gave rise to new physiological relationships between visceral and somatic sensorimotor processes. When aroused, young are channelled into particular patterns of behaviour not only by the prevailing exteroceptive stimulation, as earlier, but by intraorganic factors among which previous suckling experience plays a crucial role.

## Huddling and Related Behaviour

Newborn huddle from birth onward : after nursing has been completed and the mother leaves the nest or home site the young crawl into contact with one another and form a closely knit huddle. Since heat loss is dependent upon body surface exposure to the environment in young with poor thermoregulation, the huddle maintains body temperature and metabolic activity. Huddling is an early form of social behaviour among littermates that may be closely related in development to later forms of group activity (e.g. play).

### Huddling and Thermotactile

*Cosnier and Alberts* have shown that early huddling among rat pups is based upon thermal and tactile stimuli provided by littermates. Pups up to 10 days of age are more attracted to a warm live pup (i.e. 37°C) than a cool dead one (25°C) but the warm pup can be replaced by warm tubing or a warm fur-lined nylon sleeve. Pups are particularly stimulated to huddle when the surrounding floor is cool. *James, Welker and Crighton & Pownall* analysed the behaviour of individual puppies, when separated from their littermates in warm and cool environments, in an attempt to understand the thermotactile basis of huddling. Separation stimulates pivoting and vigorous crawling until the puppy makes head contact with its littermates.

The essential feature of the contact is thermal, however, since contact with cooled puppies or contact with littermates in an overly warm environment elicits withdrawal rather than huddling. Within the huddle temperature is maintained at 30°C and puppies remain quiescent, breathing quietly and in a relaxed sleeping posture. Warming the huddle to 32.5°C causes dispersal as each puppy responds to contact with its neighbour by withdrawing; cooling the huddle intensifies huddling as each puppy responds to the warmth of its neighbour by approaching it. The greater thermotactile sensitivity of the face and snout over the remainder of the body assures that puppies face into the huddle. The few observations of kittens indicate that similar factors operate. The role of tactile stimuli in huddling receives some confirmation from a study by *Scott et al.* (1974) in which puppies that were isolated in a warm environment but nevertheless vocalised were induced to become quite by confirming them in a section of stove pipe the floor of which was lines with soft material. The tactile stimulation, and perhaps the insulation provided, are similar to that provided by neighbouring puppies during huddling.

### Huddling and Olfaction

*Singh & Tobach* (1974) found that bulbectomized rat pups, from the second day onward, tended to be apart from the little for longer periods than their normal littermates, although in general their social behaviour was not particularly deviant. The suggestion that olfaction plays a role in huddling was more clearly established by *Alberts* (1974). Using intranasal infusion of zinc sulphate to make pups anosmic he found that huddling was completely eliminated in pups 10 days age and older, but was retained by five day old pups. *Fox* has shown that olfaction may begin to play a role in the puppy's response to its littermates as early as the first week after birth.

An odour applied to the mother's body, and therefore spread to the littermates, elicited strong approach response in puppies exposed to the odour during the first five days of life, but strong avoidance responses from puppies presented with odour for the first time on the sixth day. Similarly, among rabbits, material odour, which is shared by the littermates, elicits a strong searching response when held close to the young's nose without tactile contact as early as the second to fifth day after birth. Evidence suggests also that hamster pups may begin to respond to litter odours at the end of the first week.

### Orientation to Littermates and Vision

Huddles become somewhat less compact as young develop thermoregulation and less in need of insulation from heat loss. Nevertheless, they continue to aggregate in groups which may consist of several young instead of the entire litter as earlier. Of greater interest is the fact that they orient to littermates from a distance suggesting that vision is beginning to plat a role. As late as 17 to 20 days of age, about a week after eye-opening, puppies still do not orient to littermates in the home cage from a distance. This does not mean that they do not react to them but they continue to show circular movements similar to those seen earlier. Starting around 25 days of age, however, they begin to approach littermates from a distance directly.

*Rheingold and Eckerman* showed that kittens respond to littermates at a distance even when they do not orient directly to them. They studied the earliest age at which kittens would be comfortable by the presence of a littermate in a strange environment. At two weeks of age, the earliest age at which testing was done, kittens emitted 50% fewer vocalisation when a littermate was present compared with when it was absent. Despite the fact that only sketchy evidence is available to trace the development of huddling and other relationships with littermates among these altricial species, the main outlines of this development are clear.

Stages in the development of responses to littermates are similar to those found in the development of suckling : there is an initial stage of dependence upon thermotactile stimulation, a later stage of olfaction based responses and a final stage of visually guided responses to littermates. Moreover, early responses to littermates are dominated by exteroceptive stimuli while later stages show evidence of an increase in self-initiated approaches to siblings based upon the developing social relationships. Huddling is a relatively simple behaviour pattern compared to suckling, but it represents only the earliest form of interaction among littermates that ultimately develops into the complex patterns of play.

## Development of Home Orientation

The nest or home site is the socially conditioned environment in which the young undergo the early stages of their development. The young, from an early age, develop a pattern of home orientation that plays an important role in their relationship to the mother, and the organisation of their behaviour in relation to their surroundings and littermates. Since the original discovery of home orientation in the kitten orientation to the site has been found in the hamster, rat, and puppy, in one form or another, suggesting that further research will reveal that this is a widespread and basic phenomenon among altricial newborn.

## Thermotactile Basis of Home Orientation

In all altricial species the nest or home site usually provided with a soft insulating material and is warmer than the surrounding regions, largely as a result of warmth from the mother's body, the huddling young, and the insulation. *Tobach et al.,* found that three day old rat pups were more active in shavings, the usual nest material used in laboratories, than on the surrounding bare floor, on which they remained immobilised. The soft material used to line the nest site in many species is likely to be more attractive to newborns than coarse material because of its tactile properties, as the studies on newborn puppies by *Stanley et al.*, have shown. While these tactile properties are attractive they are of little use in orientating to the nest from a distance; they function mainly to keep the newborn in the nest or to bring to rest a newborn that has been wandering. The situation with respect to the thermal stimulation provided by the nest or home site is quite different.

*N. Freeman* measured the floor temperature at various regions of the home cage of a mother cat and litter just after they were removed in preparation for testing kittens in home orientation. The floor temperature of the home site was 35.3°C at its centre, 33.6°C at its border, about 25 cm away from the centre, and it decreased from 25.2 to 23°C at a distance of 15 cm. This thermal gradient persisted for some time after removal of the litter, but, of course, if the mother and litter remained in the home region it would be maintained constantly. On the basis of the thermal gradient kittens could distinguish the home from the adjacent corner when placed there during a test. They vocalised loudly and moved about actively in the adjacent corner while they soon became quiet and fell asleep when placed in the home site.

Odours are not involved in this response initially, since the thermal gradient produced by a heat lamp on a floor surface that had not been occupied by the mother and litter or any other cats had the same effect. The existence of the thermal gradient between the home site and the adjacent corner also enabled kittens of a slightly older age to crawl to the home from this outlying region. Also, when tested on a thermal gradient ranging from 15.6°C to 48.3°C over a distance of 56 cm, kittens up to the age of one week or so, crawled in the direction of the higher temperature if they were placed at a region below their preferred temperature. Similarly, starting at one day of age and continuing through the eighth or ninth day, hamster pups orient to the warm end of a thermal gradient, crawling distances as great as 45 cm in one minute.

*Okon* has shown that hamster pups are distressed in an ambient temperature of 28°C and emit audible and ultrasonic vocalisations. They also become responsive to a thermal gradient and crawl rapidly to the region of their preferred temperature of 33°C. The age at which thermal orientation wanes in hamster pups is variable and depends upon ambient temperature. As pups become older (i.e. 11 days and older) cooler temperatures are required to stimulate thermal orientation: at warm ambient temperatures, still below their preferred temperature, thermal orientation may wane as early as the sixth day. It is clear that thermal orientation is based in part upon a contrast between the pup's body temperature and the thermal gradient and also between the floor surface under the pup and the floor temperature of a neighbouring region. In rat pups the ability to orient to a thermal source develops over the first two weeks with an important change occurring around the end of the first week.

Until the age pups react to contact with a thermal source by remaining in contact with it but they do not orient to it from a distance. This is in agreement with Cosnier's findings, based upon tests in which pups were placed directly in contact with the heat source. Between birth and the fifth day, pups become increasingly active when the ambient temperature is reduced. If they make contact with a warm area they come to rest but there is no indication that they orient to the warm area from a distance. *Alberts* reported that wandering in response to cool ambient temperatures was the basis of huddling among five day old pups and *Fowler & Kellogg* have confirmed this.

After the fifth day, pups will orient toward a heart source provided they can detect it from a distance through a thermal gradient. *Fowler & Kellogg* found that during the first five days pups exposed to an ambient temperature of 21-23°C for one hour declined in body temperature to between 28 to 29°C, indicating failure of thermoregulation with respect to the usual body temperature of pups at an older age (i.e. 36-37°C). Although body temperature decline during this early period, thermal orientation was not evident until after the fifth day. To test thermal orientations *Fowler and Kellogg* (1975) placed rat pups between two chambers only one of which was warned. Between the starting point and the warmed chamber, located about 15 cm away, floor temperature increased from 23 to 36-37°C with a 1°C increase as close as 5 cm from the starting point. The unheated chamber was at 22°C, the ambient temperature during testing. Thus, in effect, the pups followed a rather steep thermal gradient (i.e. 14°C over 14 cm) in crawling from the starting point to the warmed chamber.

In a less steep thermal gradient pups are able to turn in the direction of the warmed floor surface but are unable to follow it for any appreciable distance. Thus orientation to the nest or home site arises early in development among altricial newborn : they respond to lowered temperature (within limits) by calling and increased locomotion. At the earliest age they are poorly equipped to respond to thermal gradients but they do respond to entering or being placed directly in a warmed region: they become calmed, vocalisation ceases, they come to rest and usually fall asleep. The activation caused by cooling is reduced by warmth. At a slightly older age they respond to cooling by adopting a path along a thermal gradient, leading to the warmer region. This ability improves over the next period but, as we shall see, it often gives way to orientation based upon olfaction. However, the calming effect of warm temperatures remain an important feature of the nest of home site beyond the period when orientation to these regions is based upon thermal gradients.

**Home Orientation and Olfaction**

The nest or home site differs from the surrounding regions in the odours deposited there by the mother and litter. These too provide a basis for orientation to the home site. Olfactory orientation to the home site develops after thermotactile orientation has been established and it may be based upon this earlier form of orientation. Towards the end of the first week, a warmed floor surface is not sufficient to quite kittens and a thermal gradient between two cage regions, lacking home cage odours, does not result in kittens adopting a path from the cooler to the warmer region.

Only the presence of home cage odours quiets kittens and only if the home site has the usual odours and, presumably, an olfactory gradient, can kittens find their way back to it. From the seventh day onward, home orientation in kittens is based mainly upon olfaction, although at the home site, the combination of olfactory and thermal stimuli is more effective than either alone. In the hamster thermal orientation declines as orientation based upon odours from the nest becomes established. Starting around the seventh and eight day, hamster pups turn toward nest shavings and away from fresh shavings when placed on a screen above the border between the shavings. They spend 80% to 100% of their time crawling in the nest shavings. Shavings from nests inhabited by a mother and litter are preferred to shavings from nests inhabited by either a male or a non-pregnant female, and nest shavings from four day old litters are as effective as shavings from older litters.

Earlier, hamster pups can distinguish between odours, preferring, for example, the odour of cedar to pine shavings; this appear on the third day.

Nest odours must require an additional period of four or five days to become associated with thermotactile stimuli from the mother and littermate to provide a basis for orientation to the nest. As indicated earlier, rabbit pups and puppies begin to respond to odours from the mother before the end of the first week and it is likely that the same odours deposited in the nest would provide a basis for home orientation. It is clear that the odour acquires its attractiveness through association with other stimuli from the mother during the first five days. Home orientation in puppies that appears to be based upon olfaction has recently been reported by *Scott et al.* Puppies were placed either in their own nest boxes, emptied of the mother and littermates, or in a similarly constructed nest box that was lined with fresh towelling; vocalisation were recorded in each.

One precaution was taken, which, however, may have affected the results: to exclude the possibility that puppies might respond to the thermal difference between the litter nest box and the unused one, both boxes were warmed to about 29°C before testing. Despite this, vocalisations were more frequently in the unused nest box than in the litter nest box from around the sixth to eight day in several litters and on the eleventh-twelfth day in all litters. Vocalisations were infrequent in both conditions at first but they rose sharply after the twelfth day in the unused nest box. The distress caused in puppies by placing them in a strange nest box is similar to the distress caused in kittens by placing them in a strange cage: the rate and intensity of vocalisations rises during tests, and with age they become more frequent and more intense.

The locomotory behaviour of the puppies in the above study was not described but, in an earlier paper with older puppies, vocalisation was accompanied by an increase in activity in a strange pen and presumably the same occurred here. Rat pups exhibit the first signs of orientation to the home sage from a distance of less than 30 cm on the third day and even more clearly on the fifth and sixth days. Placed on a 11.4 cm diameter circular platform, located midway between the home cage and a strange cage, pups point themselves toward the home during most of a three minute test but they are not yet able to crawl to it. Between the ninth and twelfth days pups become better able to locomote over long distances by crawling. On the ninth day, pups placed in a neighbouring cage reach the home cage in 50% of the tests by passing through a narrow alley, with latencies ranging from two to three minutes. Latencies to traverse the 15 to 30 cm distance between the starting chamber and the home cage became shorter on the twelfth day (i.e., one to two minutes) and 85% of the pups are successful and on the sixteenth day, around the time of eye-opening, nearly all pups orient to the home with latencies ranging from 30 to 40 seconds. At 19 and 21 days latencies are 10 seconds or shorter. Neither the paths taken to the home cage nor the sensory cues utilised by pups were analysed in the above studies.

*Turkewitz* reported that successful orientation from a neighbouring cage to the nest site in the home cage by 9 to 12 day old rat pups was accomplished indirectly through wall-hugging. The long latencies to reach the home cage in the above studies suggests that a similar mode of home orientation occurred at 12 days of age. Old pups, however, appear to adopt direct paths to the home cage and as a consequence latencies are considerable shorter. On the basis of *Turkewitz's* study, it would appear that olfactory and perhaps thermal stimuli are involved in home orientation by rat pups before their eyes open on the fifteenth to seventeenth day. The long latencies suggest pups adopt an indirect path to the home and this is more characteristic of young that orient on a non-visual basis: the shorter latencies after eye-opening certainly contrast with the longer latencies earlier. Findings reported earlier, that rat pups show evidence of olfactory discrimination of nest shavings as against fresh shavings around the ninth day and respond to maternal odours by the fourteenth day, would also support this interpretation of the above home orientation.

## Home Orientation and Vision

The onset of vision marks the beginning of the decline of home orientation among altricial young. With the ability to view the home from a distance it is no longer necessary to return to it: instead young tend to remain in the vicinity of the home or nest site for a period before they disperse. Moreover, orientation to the home declines also because the young begin to follow the mother and littermates and these became centres of orientation rather that the nest or home site. There is, however, a short period after eye-opening and the beginning of vision when home orientation continues and vision is used to enable the young to return to the home from a distance. In the rat, as we have seen, the latency for reaching the home from a nearby region is reduced to a few seconds shortly after eye-opening. This indicates that the young begin to take direct routes to the home, and this probably depends on visual stimulation.

The hamster pup shows a decline in olfaction based orientation beginning around the thirteenth day, shortly after eye-opening, and it is completely absent by the nineteenth day. This may be based upon the gradual increase in visually based orientation. Among kittens the use of vision in home orientation begins around the fourteenth day. Testing kittens in the dark results in some decrement in performance with indications that the kittens are seeking the visual stimuli to which they are accustomed. However, they are still capable of using olfaction and most of them reach home. If olfactory cues are removed, some kittens are still able to reach the home and this increases after the seventeenth day. The use of vision in home orientation among kittens eventually leads to the decline of this behaviour at around the eighteenth to twenty-first day. Instead of returning to the home region, kittens look forward it, or they may actually enter the home region briefly then leave it.

The ability to see the home often enables kittens to wander more freely around the entire cage since the home region is always in view. However, when the home was made especially prominent visually by placing a fur piece there, home orientation persisted until 45 days of age. With home odours absent, kittens began to show this visually based home orientation around 25 days of age gradually perfected it, traversing the age, a distance of about 76 cm, in less than 20 seconds as they moved directly toward the home and came to rest entering it.

## Behaviour Development of Altricial Young

The early behaviour development of altricial young can be divided into three stages: in the following section we shall analyse these stages and the transitions from one stage to the next.

### First Stage: Thermotactile Stimulation

The neonates' earliest behaviour is organised predominantly in relation to low intensity thermotactile stimulation provided by the mother, littermates and nest or home site. These elicit from the newborn either general or specialised approach responses. The general response of forward crawling is elicited by centrally applied head and face stimulation over a broad area and turning is elicited by laterally applied stimulation, through the close relationship that exists between anterior end stimulation and movement of the forelimbs in an alternating paddling motion and the hind-limbs in a joint pushing action. Specialised responses are elicited by localisation stimulation of the snout, mouth, and lips: the specialised responses take their character from the peripherally organised pattern of movement characteristic of the local region and the restricted locus of stimulation. The thermal and tactile characteristics of the mother, littermates and nest or home site provided patterned thermotactile stimuli which elicit from newborn the general and special approach responses necessary for the initiation of suckling, and huddling, and for remaining within the confines of the home. These responses are only loosely patterned sequences at first, depending upon kinds of stimulation the newborn encounters in succession as it crawls forward.

Stimulated initially to approach the mother by contact with the attractive thermotactile stimulation of her body, the newborn subsequently encounters the more detailed stimuli of the mother's fur around the nipple, the areola, and finally the nipple itself. At this early of development the central control of behaviour is not yet strongly developed beyond the mediation of the admittedly complex sensorimotor integrations underlying general and specific approach and withdrawal responses to stimulation. It is doubtful whether, for at least a short time after birth, the functionally distinct behavioural responses to the mother, littermates and nest or home site are represented by equally distinct central regulatory process. The underlying similarities between the behavioural adaptations to the mother, littermates, and nest or home site, arise from the fact that they share in common responses to similar thermotactile stimuli.

The observation differences arise from the fact that in each of these responses the neonate also responds to specific stimulus properties of the mother, littermates, and the nest or home site. While the relative contributions of central regulatory process and peripherally elicited responses favour the latter in the neonate (as compared with older animals in which central regulation plays the dominant role) central processes are not without effect on the neonate's behaviour. Exteroceptive stimuli (i.e., thermotactile) appear to have two concurrent effects upon the neonate. They arouse the neonate to activity, as when the mother's licking activities the sleeping neonate to raise its head and begin to crawl. Once the newborn is aroused, stimuli also elicit approach responses of a general and specialised nature.

Thermotactile stimuli are therefore both arousal inducting and response-eliciting. Neonates that have been aroused are more sensitive to exteroceptive stimuli that follow and exhibit more vigorous responses to these stimuli. The central component of exteroceptive stimulation therefore is an arousal that in turn potentiates the responsiveness of peripheral processes. Early in development central arousal is relatively non-specific in its afferent and efferent relationship and this contrasts with the specificity of the peripheral responses that are elicited by thermotactile stimuli. It is this contrast between the greater specificity of peripheral responses–their direct relationship to thermotactile stimuli–and the non-specific character of arousal and its contribution to early behaviour, that has led many to characterise the neonate's behaviour with some justification as simply reflexive in nature. Early suckling among newborn rat pups and puppies exemplifies the relationship between central and peripheral processes in the regulation of feeding.

The compelling effect of thermotactile nipple stimuli in eliciting sucking is shown by the fact that prior feeding or stomach loading cannot prevent the young from grasping the nipple and sucking. In the rat pup, if milk is not forthcoming sucking may continue indefinitely; suckling does not continue indefinitely in the kitten but its duration is not shorter that of kittens that obtain milk. The difference may rectify the fact that the rat pups tested with anaesthetised mothers while the kittens were tested with awake, non-lactating females. Yet normally rat pups and kittens do not suckle interminably once they attach to nipples. After a period they began to loosen their grip on the nipples and slide off, or they are detached from them when the mother rises to leave.

The intake of milk has an effect on the peripherally organised response of sucking, but the effect appears to be mediated by a lowered arousal, to the point of sleep. All overt activity ceases, not only suckling, and newborn's behaviour alternates between sleep and feeding for a considerable period after birth. Preloading of newborn with milk has an effect on suckling when the amount preloaded is sufficient to induce sleep. With effort, even under these conditions, suckling can be elicited by bringing the newborn's mouth to the nipple but the central effect of intraorganic stimulation arising from the distended stomach or sensory receptors in the stomach, usually prevails and the newborn releases the nipple soon after grasping it. Preloading also reduces milk intake but the basis for this is not clear. Varying levels of central arousal appear to influence the responsiveness of newborns to exteroceptive stimulation: low-level arousal, that characteristic of newborn shortly after feeding dampens the effect of exteroceptive stimulation, while very high levels of arousal, following long periods without food or upon exposure to low ambient temperature, produces an excessive response to exteroceptive stimulation.

There is therefore an optimal level of central arousal at which the newborn exhibits its typical responses to exteroceptive stimuli. The source of central arousal in the newborn are interoceptive as well as exteroceptive, as studies on stomach preloading indicate. These sources have not been studied to any great extent, but interoceptive stimuli have a history in prenatal ontogeny that predates exteroceptive stimuli. Exteroceptive stimuli may in fact exert their influence, in part, through their effect upon interoceptive processes, as for example in responses to ambient thermal stimulation. During the earliest postnatal stage there is a close relationship between those exteroceptive stimuli (i.e., thermotactile) to which newborn are most responsive in their behaviour toward the mother, littermates, and nest or home site, and the conditions which stimulate it to activity. Thus, newborns are stimulated by loss of contact with the mother or littermates and by displacement from the nest or home site; in the latter cases it is probably exposure to lower ambient temperatures than are normally present in these situations which stimulates, the newborn as studies on kittens have show.

At this early age many different objects having attractive thermotactile properties may be equivalent in their arousal and calming effect upon newborn because the special properties of the mother, littermates and nest or home site do not yet play a role. This fact has allowed investigators to ignore the social nature of the newborn's responses to species mates and to the socially conditioned home and to introduce special criteria for when the young's responses are to be considered social. Since the usual criteria are based upon visual responses to species mates, responses that appear during the third stage of early development are more likely to be called social responses and social behaviour is therefore viewed as a later development. Such a view ignores the ontogenetic basis of social responses and their relationship at each stage to the behavioural capacities and behavioural organisation of the young. The earliest changes in the newborn's behaviour are the formation of extended action patterns, incorporating the earlier hesitating and variable crawling approach response and the specialised feeding and huddling responses into more smoothly co-ordinated patterns.

In brooder-reared kittens, after a day or two of feeding, the kittens crawls to the correctly textured flange (i.e., that association with a nipple that gives milk) in a smoother more patterned movement, with short pauses at crucial points such as the edge between the brooder and the floor and between the brooder and cover and the textured flange. At there points the kitten sniffs the brooder cover and rubs its snout against its surface as it does when it reaches the flange. A short period of flange contact is followed by nipple localisation and nipple grasping and sucking. Among puppies approached to the nipple can be associated with either soft-or hard-textured paths with gradual improvement in turning toward one or the other and heading rapidly for the nipple, grasping it and sucking.

Repeated experience with either warm or cool thermal stimuli along alleys leading to a nipple results in a gradually more rapid crawling approach to either. The formation of these specialised action patterns based upon thermotactile stimuli indicate progress along several lines: approach responses become specialised in relation to the young's discrimination among thermotactile stimuli. In this respect approach responses are formed more rapidly when a low intensity, approach-eliciting stimulus is the positive one than when a stimulus that is initially either weakly approach- eliciting, or is actually withdrawal-eliciting, is used as the positive stimulus. In addition, components of the approach response are integrated into a pattern that appears to be less dependent upon continuous guidance by exteroceptive stimulation and more dependent upon central regulation.

In suckling approaches, striving toward the nipple indicated an early appearance of anticipatory responses in advance of actual contact with the nipple region. These changes channel the earlier general arousal along specific lines. Thermotactile stimuli are no longer equally effective in eliciting approach responses: the newborn turns away from certain thermal stimuli and becomes highly excited when it makes contact with others. There are indicators in the study of *Bacon (1974)* that negative stimuli do not acquire inhibitory effects at this stage but rather that positive stimuli acquire heightened arousal effects. These may have their origin in the heightened arousal that occurs when the young are stimulated by milk sucking. At this stage, new sources of disturbance may arise when the usual incentive is absent or is altered; this is in fact evidence that action patterns of a broader nature already exist in the young and that they are highly specific to the stimulus conditions in which they were formed.

**Second Stage: Olfactory Stimulation**

Although in altricial young the second stage of early behaviour development is characterised by the growing influence of olfactory stimulation, its main feature is the increasing specificity of behavioural responses in relation to the familiar objects in the environment. The young use olfactory stimuli to differentiate between familiar and unfamiliar object. The initial exposure to olfactory stimuli occurs during responses to thermotactile stimuli, but in the beginning olfactory stimuli can provide little guidance to the newborn. Except when it encounters strong aversive olfactory stimuli, which induce withdrawal responses, olfactory stimuli cannot elicit either general approach or specialised responses in the newborn. Not until responses have begun to become organised into specific action patterns based upon thermotactile discrimination, and the central arousal processes are channelled, can olfaction begin to play a role. The suggestion is that olfaction advances further the process of discrimination among objects, along lines that are necessarily specific to the mother, littermates and nest or home site.

More important, however, is the ability of olfactory stimuli to arouse the initiation of action patterns that are then guided by combined olfactory and thermotactile stimuli. Olfactory stimuli may therefore determine whether or not a response will occur. There is associated with the onset of olfaction the appearance of olfactory orientation behaviour by means of which young explore their olfactory environment. *Welker* has described the development of sniffing as an olfactory exploration pattern in the young rat, and *Komisaruk* has added important details to the analysis of olfactory exploratory behaviour in this species. When placed in a new environment young initiate non-specific olfactory exploratory behaviour which soon gives way either to specific behaviour patterns upon identification of the odour or to further tactile exploration with the vibrissa. These exploratory patterns range from air sniffing to sniffing of the floor surface and approach and sniffing of objects in great detail. Thus, for example, by sniffing the floor surface kittens find their way to the home region and rat pups by sniffing the air are able to locate the mother at a distance of more than 30 cm. Depending upon the distribution of odours, therefore, olfaction enables young to develop distance perception of objects and to initiate movements towards objects before actual contact is made with their thermotactile properties. In this sense, therefore, olfaction plays a large role in developing central control of action patterns begun with respect to thermotactile stimuli.

Anticipation of forthcoming stimulation and the associated action pattern, exhibited during the latter phase of the first stage of development, is gradually transformed into self-initiated approaches to objects with anticipatory action patterns almost entirely centrally aroused, ready to be performed when the object is reached. Rat pups deprived of olfaction during this phase appear much less oriented to significant social stimuli and to the nest site (*Singh & Tobach*, 1974). The stage is set during this second stage of early behavioural development for the incorporation of vision into the central control of action patterns during the third stage of early development. As olfaction begins to contribute to central arousal processes and the action patterns to which they give rise, it begins to play an increasing role as a motivating condition and as a possible source of distress.

Rat and hamster pups, kittens and puppies begin to show distress when removed from their familiar olfactory environment. In kittens and puppies this is evident at the end of the first week and it appears at a slightly older age in rat and hamster pups; it provides the basis for these young to initiate movement toward littermates, resulting in huddling, and toward the home or nest site during the development of home orientation. Moreover, olfaction serves as a goal for these behaviours in the sense that upon reaching the familiar olfactory situation in the huddle or in the home region, young are calmed and soon come to rest. It is apparent, therefore, that not only perceptual and motor processes become more complex as development proceeds but motivational process share in this growing complexity.

**Third Stage: Visual Stimulation**

Vision greatly enlarges the young's capacity to differentiate perceptually between the significant objects in its environment. Neither thermotactile nor olfactory stimuli are able to convey to specific location of their nest or home site. While olfaction may indicate that the mother is nearby it does not indicate what she is doing, while vision conveys both. This stage is marked by an acceleration in the young's development of social interactions based upon vision. Initially social interactions consist of visual approaches to species mates at a distance, at which point action patterns based upon olfactory and thermotactile stimuli take over : vision therefore contributes little to the character of the interactions bur does contribute to their occurrence. Thus, for example, kittens at three weeks of age, approach the mother at a distance for suckling but when they reach her, they adopt their earlier mode of nipple searching with their eyes closed.

Gradually, however, vision comes to trigger not only the approach to the mother, but also those action patterns that were formerly triggered by non-visual sensory systems. Kittens at this stage, approach the mother and reach up from beneath her as she remains standing, locating and grasping a nipple without any significant preliminary nuzzling. Brooder-reared kittens walk directly to the nipple and grasp it instead of following a path of nuzzling on the brooder surface, and cage-reared kittens walk directly across the cage from the diagonal corner to the home region instead of crawling along a path that passes through the adjacent corner. Vision accelerates the process of increasing central regulation of behaviour and with this furthers self-initiated behaviour with well-defined goals.

Perceptual differentiation and the multiplication of action patterns, and their growing complexity, imply higher specific central stages of arousal and complex interactions, both facilitating and inhibitory, among these different stages. During the period of transition from olfactory to visual control of behaviour in the young, interoceptive stimulation appears also to become more influential and specific in its effect upon the central state of arousal. At this age, *Hall* found that food-deprived and non-food-deprived rat pups began to differ in their suckling behaviour, the former being faster to initiate suckling with an anaesthetised mother and abandoning suckling if no milk was forthcoming.

Similar findings have been reported in three four week old puppies and kittens whose suckling approach to the mother was influenced by stomach preloading with relatively small amounts of milk. The goal-directed character of behaviour at this stage is exemplified by the persistence kittens show in following the mother around the cage, while they look for an opportunity to suckle. At the slightest pause, they immediately reach up to her nipple and if they are detected by her movement they resume following her.

At other times, when tired, they walk across the cage to join another kitten that has settled in a corner to sleep and huddle against it, falling asleep. While they remain highly responsive to environmental stimulation, their behaviour is not directly elicited by this stimulation: rather, they are capable of making perceptual discriminations among the various stimuli and which stimuli they respond to depends upon the central state operative at the time. During this third stage of development kittens and puppies that are placed in novel environments exhibit signs of distress (i.e., vocalisation and agitated movement) that are relieved when littermates or the mother are placed in the same environment.

Rat and hamster pups orient to the mother and littermates when displaced to a neighbouring, strange cage. Thus, the absence of familiar visual stimuli becomes a source of disturbance that motivates young to region visual contact with social companions. Often the experiment serves a similar role if he has become familiar to the young. Since visual stimuli are likely to be a principal source of social stimuli from this age in into adulthood, and social responses of a clearly recognisable nature are particularly evident during this third stage, this stage had been singled out by *Scott* as the beginning of true socialisation. This analysis has shown, however, that social responsiveness arises almost immediately after birth and continues to grow in depth and complexity during each stage of ontogeny. Visually based social responses have their ontogenetic origin in earlier non-visual stages and they retain their close relationship to these earlier stages throughout life.

**Functional Aspects**

The neonate's dependence upon thermal and tactile stimulation in its behavioural responses to its surroundings is based upon its needs to maintain an optimal thermal environment for adequate physiological functioning in the face of its inability to regulate fully its own body temperature. Contact with the mother" body causes a rise in body temperature and huddling with the littermates, shown by *Albert* to result in minimal exposure of the group to ambient thermal conditions, maintains body temperature, and reduces metabolic activity. *Leonard* has suggested that overall rate of heat loss is the stimulus to which newborn hamsters respond in the thermotaxic orientation, and the effect of prolonged exposure to low ambient temperature on ultrasonic and audible sound emission by rat and hamster pups neonates respond to altered thermal conditions, particularly during nipple searching, indicates that thermal sensory receptors located in the snout, lips and face also play an important role. Neonates whose behaviour is organised in relation to thermotactile stimulation are necessarily confined to the close proximity of eat sources and their tactile representatives, and they must possess means of reaching these heat sources from short distances if they are displaced.

The behaviour of neonates is therefore in spatial scope and in the range of thermal conditions under which they can function adequately. As these thermal limitations are relaxed with the development of thermo-regulatory mechanisms, neonates become capable of extending the scope of their functioning, both spatially and with respect to environmental conditions. Olfactory stimuli play an important role in this process. They arise from substances deposited by the mother in the vicinity of the home or nest and have special significance for the newborn, since they have been experienced in conjunction with thermotactile stimulation from the mother, and therefore they can provide means for the newborn to extend its scope of activity. The mother creates an olfactory zone around the nest or home (as well as in the home itself when she is absent) in which the newborn can function because of its growing responsiveness to olfactory stimulation.

Moreover, odours are species and individual-species: they provide a basis for the specialisation of responses to particular kinds of animals and particularly to the individual mother and littermates. Evidence of olfaction-based individual attachments by neonates have been reported for rats and kittens and will, very likely, be found in many different altricial species. Home or nest orientation develops in relation to the differential distribution of odours in the vicinity of the nest or home site. It requires a responsiveness to gradients of olfactory stimulation or to orientated deposits which are the products of the mother's own position in relation to the home or nest as the centre of her maternal activity. The maturation of vision increases even further the scope of the young animal's activity and the period of combined olfactory and visual functioning ensures that early visual functioning will be in relation to the most significant features of the young's social environment and within the zone of its socially conditioned physical environment.

The specialisation of the young's responses to qualitative features of stimuli (rather than simply to quantitative features of the earlier phases) advances further with the introduction of vision. A wider range and greater variety of social signals are displayed visually than through odours or thermotactile stimuli. Play arises during the period of visual functioning and is based largely upon the young's responses to the variety of visually perceived actions on the part of siblings and other familiar contemporaries. Earlier sensory systems do not, of course, fall out of use when newer ones mature and become functional. They do, however, play a different role in behaviour than during the phase of their predominance.

## Play Behaviour

Play behaviour has been defined and characterised in many different ways. *Fagen* defines play as "an inexact term used to denote certain locomotor, manipulative and social behaviours characteristic of young (and some adult) mammals and birds under certain conditions in certain environments." We can characterise play in various animals and the groups according to the actions of the animals and the contexts in which the play behaviour is observed. Most investigators recognise at least three types of play, with some overlap. The first type is social play as exemplified by wrestling, chasing, and tumbling activities of the young of many species. A second type provides exercise for developing muscles, locomotor patterns, and other movements. The third type is often labelled "diversive exploration."

Generally, this type of play involves sensory inspection of an object followed by extensive repeated manipulation of the object. Play behaviour of one or more of these types has been recorded in mammals, ranging from rodents and bats to bears, cats, elephants, and whales. Play has also been recorded from a large number of avian species, including raptors, passerines, aquatic or oceanic birds and parrots. To date only a few anecdotal bits of evidence exist regarding play behaviour in other vertebrate groups or among invertebrates. However, additional observations must be made before reaching any firm conclusions limiting play to birds and mammals.

**Functions**

Investigators have ascribed a wide range of functions to play behaviour. In the most general sense, play functions as practice for adult activities. Animals perform many actions in the course of play which contain elements of behaviours seen in later adult life. Perhaps aggressive behaviour is the best example. As young cats or primates engaging in various forms of social play, their mock attacks, chases, and mild, non-injurious bites are all partly practices for *real life* use of these same patterns a few months or years later. A second function often ascribed to play behaviour is to aid in the process of maturation- growth and development. As young foals or lambs cavort about, alone or in small groups, they use their muscles and develop co-ordinated movements. A third function of play, may be to gain information about the environment; this would be particularly true of diversive play. By exploring and manipulating objects found in their environment, young animals accumulate information that may prove useful later in life.

Finally, play can function in the establishment of social relationships with peers and adults. Some of these may be purely affiliations, whereas others could be related to the establishment of dominance-subordinance relationships. For many primate species and some canids, the aggressive play observed in young juveniles gradually become more intense and adult-like. The patterns of dominance established in play encounters as juveniles can be retained in later adult life.

**Canids**

One animal group for which play behaviour has been studied extensively is the canids *Bekoff* studied social play and play soliciting in coyotes (*Canis latrans*), wolves (*C. lupsus*), and beagles (*C. familiaris*). His observations indicated some species differences and some age-specific trends for several aspects of play behaviour, for example, play soliciting and agonistic behaviour. *Bekoff* notes that the beagles displayed an early onset for play soliciting and exhibited very little agonistic behaviour during play–on fighting occurred at all, only mild threats. Wolves showed moderate levels of play soliciting, with an increase at the last age interval recorded in the sample. Like the beagles, wolves displayed low levels of agonistic behaviour, consisting primarily of threats. In contrast, coyotes exhibited higher levels of agonistic behaviour through the observation period, and a correspondingly lower rate of play-soliciting actions.

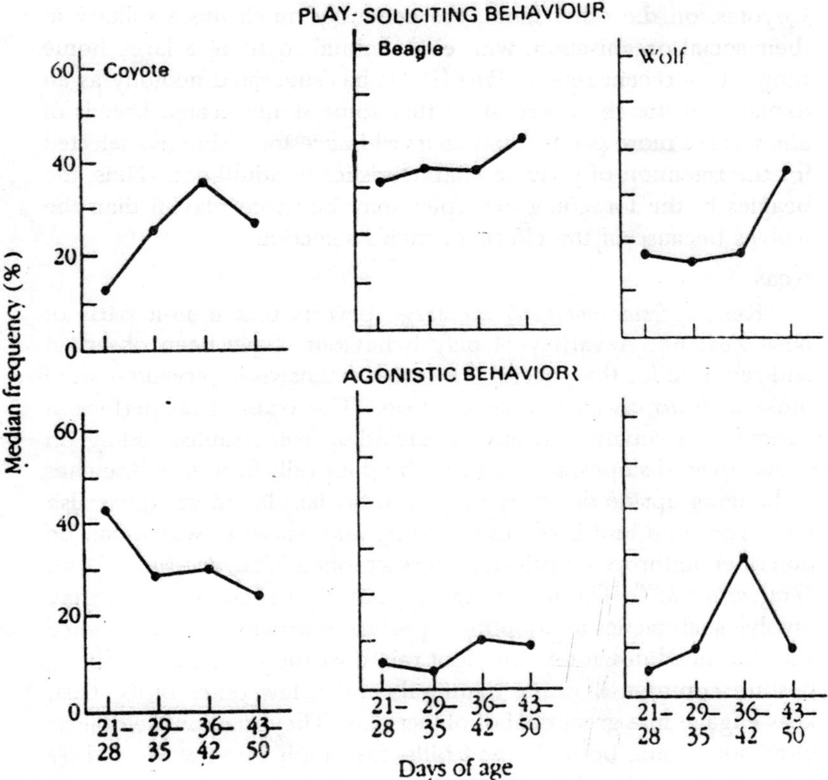


*Fig. 2.14. Play-soliciting and agonistic behaviour in canids.*

Coyotes generally establish dominance relationships through fights at an early age; this may account for the differences between these animals and the other two species. When all play behaviours were summarised, the beagles were seven times more playful than the coyotes and three times more playful than the wolves. It is interesting to speculate on the possible relationship between these differences in observed play behaviour and the differences in social structures of the species tests. Wolves are generally social, group-living animals, though some individuals may lead a solitary existence for some periods.

Beagles are at least somewhat social animals, though we rarely observe such domesticated canids in semi-natural or natural situations where their feral social structure can be fully recorded.

Coyotes, on the other hand, are generally much more solitary in their social organisation, with each animal roaming a large home range. In a recent review *Price* (1984) has suggested neotony as an explanation for the observation that some domesticated breeds of animals are more playful than their wild ancestors. Man has selected for the retention of juvenile characteristics in adulthood. Thus, the beagles in the foregoing examples may be more playful than the wolves because of the effects of man's selection.

**Keas**

Keas' (*Nestor notabilis*) are large parrots that inhibit parts of New Zealand. A variety of play behaviours have been observed and reported for this species. The most extensive observations were those of *Keller* on captive keas in zoos. The young keas perform a variety of acrobatic maneuvers, including somersaults, sliding on snow-covered slopes, and hanging by their bills from tree branches or hanging upside down by their feet. Social play in groups is also quite common and involves wrestling and activities which can be described anthropomorphically as resembling "*hide-and-go-seek*" and "*king of the hill*". The solicitation to play in a social situation may involve such tactics as adopting a posture normally used in defence with the head down and one foot raised or the stiff-legged walking posture common in some mammals and a few other birds. Last, keas engage in a great deal of object play. They manipulate objects of all sorts using both feet and bills, toss items into the air and fly at them, and they play with snow when it is present.

**Human Children**

Numerous studies have been conducted by both developmental psychologists and ethologists on play behaviour in young *Homo sapiens. Hutt and Bhanvani* conducted a follow-up study of earlier work by *Hutt* in order to attempt to make some predictions about differences in play behaviour based on longitudinal sampling. A young child confronted with a new toy will first investigate and inspect it (specific exploration) and then play with the toy (diversive exploration). Three to five-year-old nursery school children can be classified into three mutually exclusive categories with regard to their responses to a new toy: non-explorers, who visually inspect the toy, but do not handle or play with it; explorers, who thoroughly investigate the toy, but fail to do more than that with it; and, inventive explorers, who investigate the toy and then play with it in a variety of innovative ways. Do these individual differences in exploratory behaviour provide any predictive insights with respect to traits in the children at a later age?

To test this question *Hutt and Bhanvani* obtained data for about fifty children from the original sample of one hundred used to generate the three categories above. The children were seven to ten years of age at the time of the second sampling procedures. Each child was given a series of tests to measure creativity and a personally questionnaire. Each child was rated by parents and teachers on behaviour, adjustment and development. The results provided support for several suggestive conclusions–really more like hypotheses for further testing:

1. Children who has been more creative and imaginative in their early play behaviour were more likely to be divergent in creativity at the later ages; this was particularly true for boys.
2. Lack of exploring was related to later lack of curiosity and adventure in young boys, and to difficulties in personality and social adjustment in young girls.

Hutt and Bhanvani note that some of these effects may be attributed to early childhood differences between the sexes; boys are more exploratory in their play activity for a longer period of their life, and girls are socially and linguistically more advanced than boys at nursery school age. We might also note that further longitudinal follow-up testing to assess longer term effects of these early differences would be both appropriate and necessary to strengthen or extend these conclusions.

# 3

# DECISION-MAKING AND MOTIVATION

The internal changes responsible for making an animal behave differently at different times. Confronted with food, animals do not always eat, nor do they attempt to mate whenever a potential partner appears. The study of motivation ,is aimed at discovering the processes responsible for such changes. A hypothetical state of the individual organism that arouses a goal directed activity or mood, readiness for behaviour, "drive". The readiness or urge of an animal for a certain behaviour. This is determined by a number of factors, which include-external stimuli, internal stimuli and hormones. At any given time if has a certain level this value declines when the act is performed then rises again. Externally, a change in mood can be recognized by the fact that the animal responds to the same stimulus differently at different times. Much of the causal analysis of animal behaviour has been concerned with identifying the relationships between responses and external stimuli.

Many activities only occur in the presence of relevant stimuli; an animal cannot eat in the absence of food, for example. It can, however, lood for food when it is not immediately available, and such food searching is behaviour that can only be explained in terms of internal processes. A more general point is that in many of their activities, animals do not always respond in the same way to external stimuli. Whereas a rat may withdraw its foot every time it steps on a hot surface, it does not eat whenever, it is presented with food. Clearly, some internal process influences whether it responds to food. Such processes, whatever, they may be, are the stuff of motivation. If asked to provide a causal explanation of why an animal does not eat when presented with food, our most likely answer is that the animal is not hungry. This appears to provide an answer by invoking an internal process but in fact it really only describes what we have seen in different words and serves only to raise more questions. What is hunger and what are the physiological events that produce it? The essence of much research into motivation is the attempt to express concepts such an hunger in terms either of observable physiological processes or of rules which may help us to predict when a particular behaviour pattern will occur. Before going further, we should consider briefly other possible reasons we might have used to explain an animal's failure to eat. One possibility is that the animal was engaged in another, perhaps more important activity, such as seeking a mate.

The motivation of one activity is dependent on that of others. Another factor that may affect the way an animal responds to external stimuli is its stage of development. Young mammals being fed on their mother's milk do not respond to food they will eat as adults and, likewise, fully formed sexual responses generally do not appear until an animal is mature. Ontogenetic changes of this kind are not regarded as manifestations of motivational processes because they are irrevesible; mature animals do not revert to immaturity. By contrast, being hungry, thirsty or sexually aroused are states that animals can experience frequently. Like maturational changes, learned responses to stimuli are usually also considered to be distinct from motivational processes because of their typically long-lasting influence on behaviour. Particular noxious flavours may be experienced once and avoided thereafter, and an animal's mating preferences may be restricted by early experience of its parents or siblings. Such effects cannot be ignored, however, but must always be taken into account in studies of motivation; as with other aspects of behaviour, motivation is subject to maturation and is influenced by learning. Many American psychologists emphasized the motivational aspects of instinct and start with the consumption that many patterns of instinctive behaviour can be analyzed upto drive directed ago.

The attainment of which results in reduction of the drive. For example, the case of a 3 years old boy with an abnormal craving for salt. From early life he always prefered salty foods and would lick the salt of crackers rather than eat them. When he was 18 months old he discovered the salt sheckers and began eating salt by the spoonful. He learned to point to the cover cuppled to take the salt sheckered. The first world he learned was salt. It turned out that his craving for salt had kept him alived. When taken to the hospital observation and placed on a standard hospital type with a limited salt, he died within 7 days. At autopcy it was learned that he had tumours of the adrenal glands and thus lacked the hormone necessary to reabsorb salt at the kidney only by constantly replacing salt lost in his urine. He could maintain himself. Thus, it can be seen that drive is a striking towards some goal to get salt.

**Intervening Variables**

Although hunger is a sensation of which we have direct experience, it is a hypothetical construct in the analysis of animal behaviour: it is something we cannot observe or measure directly, but can only infer from variations in overt behaviour. Since hunger is invoked to explain variations in response to food it is referred to as an intervening variable, because it has some modulating effect between a stimulus and its associated activity. Discussions of motivation are littered with intervening variables, such as thirst, sexual arousal, fear and, more generally, drives of one sort or another. The question we must ask if whether such intervening variables help in understanding motivation. The basic problem with intervening variables is that any attempt to explain things in these terms involves a circular argument. We may infer that because an animal is eating voraciously, it has a high feeding drive. However, our only evidence for that assertion is the nature of the animal's behaviour, which is the very thing we are seeking to explain.

It may appear that we break this circularity if we identify factors that influence feeding drive. For example, animals generally eat more readily if they have been deprived of food, and so we could say that food deprivation increases feeding drive which stimulates feeding. We can express this effect more economically, however, by simply saying that food deprivation stimulates feeding. A hypothetical intervening variable may have some explanatory value if it enables us to simplify observed relationship between several stimulus inputs and a number of different behavioural output shows three treatments that increase the tendency to drink in rats :

water deprivation, feeding with dry food and injection with saline solution. The effect of all three treatments can be observed in three measures of drinking behaviour: the rate at which a rat presses a bar to obtain water, the volume of water it drinks, and the concentration of quinine (a distasteful substance) it will tolerate in the water it drinks. The number of relationships between these treatments and measures can be reduced from nine to six by introducing thirst as an intervening variable. Invoking a single variable, such as thirst, could therefore be useful provided that the motivational processes underlying the activity concerned constitute a single process.

But there is no a *priori* basis for such an assumption and, indeed, there is abundant evidence that it is not justified in systems that have been studied in detail. The study of drinking in rats by *Miller* (1959) provides an example of such evidence. He gave rats saline solution by means of a tube inserted into their stomachs and then recorded the three measures of drinking mentioned above the various time intervals after the treatment. One would assume that, as time passed, the rats' thirst would have increased and that,

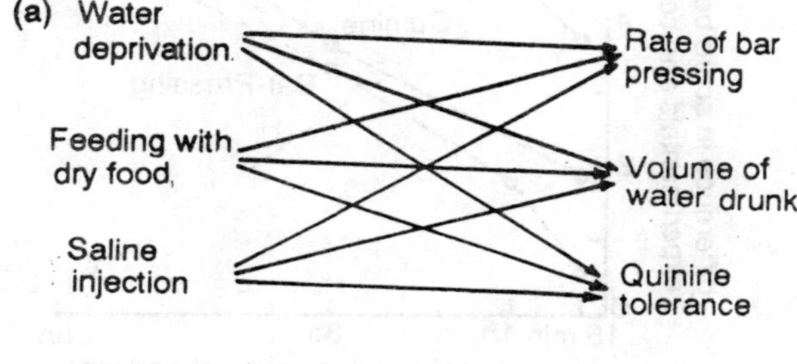


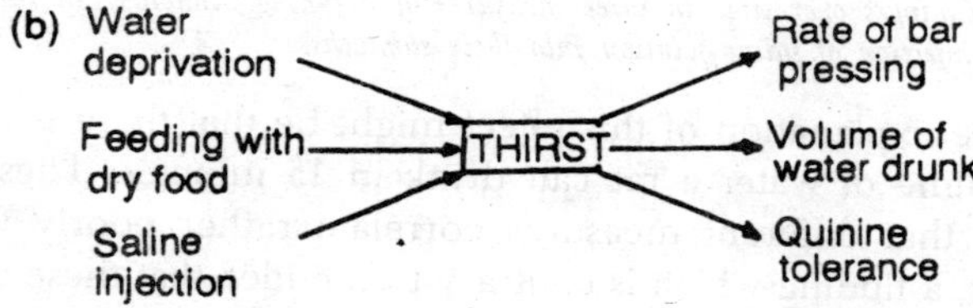


*Fig. 3.1. The role of hypothetical intervening variable in simplifying the relationships between casual factors and behavioural measures. (a) The three treatments on the left all influence the three scores of drinking behaviour on the right, in rats, giving nine casual relationships. (b) By introducing 'thirst', the number of casual relationships is reduced to six.*

if their behaviour is indeed influenced by a single intervening variable, the three measures should be closely correlated. One measure, the concentration of quinine that rats will tolerate in their water, rises at a constant rate, as one would expect if thirst increased as a simple function of time elapsed since the treatment. By contrast, another measure, the volume of water drunk in 15 minutes, shows an increase only over the first three hours and then reaches an asymptote. On this scope rats seem to be no thirstier after six hours than after three.

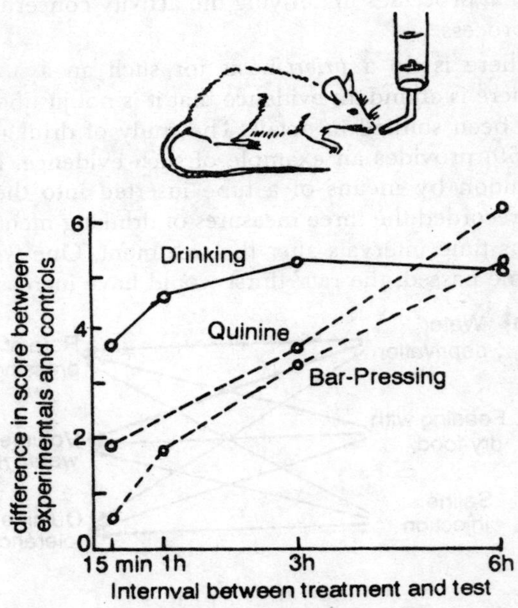


*Fig. 3.2. Changes over time in three measures of drinking behaviour in rats following injection of saline solution into their stomachs.*

One explanation of this effect might be that there is a limit to the volume of water a rat can drink in 15 minutes. These results suggest that different measures correlate rather poorly with one another, a finding which is contrary to the idea that these measures of drinking are all influenced through a single variable that we could call thirst. In summary, when an intervening variable is invoked, we should not believe that anything has been explained; all that has been done is that some phenomenon that requires explanation has been identified. We cannot even assume that a unitary process is implicated. At best, intervening variables are labels for unknown physiological processes. The ultimate aim of much research into motivation is to identify and understand how such processes work, so that concepts such as hunger, thirst and drive become unnecessary.

## A Model of Motivation

We all experience hunger at sometimes and not at others and we realise that the urge to eat is not constant. The sight of a delicious meal is on some occasions mouth-watering and on others of no interest at all. Once again we owe to Konrad Lorenz a clear theory of how such changes might come about which helps us to think about the problem. This theory was put forward in the form of the model, the word 'model' in this sense meaning a simple scheme which is proposed to work in a way similar to the real system. Lorenz's theory is known either potentiously as his 'psycho-hydraulic' model or, more prosaically, as 'Lorenz's water closet.' This model is not something one would look for inside an animal! It is what is called an 'as if' model: the animal may behave as if it had such a system for organising its behaviour within it.

The theory helps one to think about the problem and design experiments to see how the system really works rather than proposing specific mechanisms. Lorenz supposed that different actions depended for their appearance on a supply of 'action-specific energy' which accumulated with time since the animal last behaved that way and was used up as it performed the act. He visualised this as water accumulating in a tank out of which it could only escape through a value at the bottom. The valve was, however, a rather strange one. It could be opened either by the water pushing within, or by a string attached to a scale-pan pulling from outside.

To Lorenz, weights on this scale-pan were the equivalent of stimuli leading to the behaviour: the more appropriate the stimulation, the heavier the weights and the more likely the valve was to be opened. Lorenz's model has some interesting properties which can be compared with real behaviour. First, the longer since the behaviour was last performed the more action-specific energy will have accumulated and the more likely it is that the behaviour will appear. Secondly, the model suggests that the accumulation of action-specific energy will lead the behaviour to occur even if the stimuli present are slight (the weights on the scale-pan are very light).

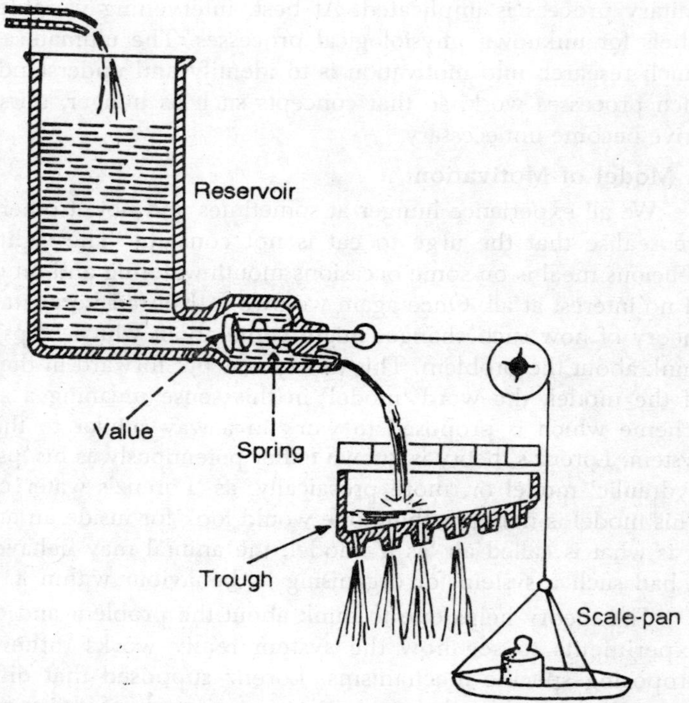


*Fig. 3.3. Lorenz's 'psychohydraulic' model of motivation. Action-specific energy is represented by water, which accumulates progressively in a reservoir when the behaviour concerned is not being expressed.*

Ultimately Lorenz argued, behaviour of which an animal has long been deprived will appear as a 'vacuum activity' with no stimulus present at all. The idea of vacuum activities came at one extreme: when there was an excess of action-specific energy. At the other was exhaustion which occurred, according to *Lorenz,* when an activity had been stimulated so often that the animal had run out of this energy. Then, there is a final idea incorporated in the model. The way the valve words suggests a particular relationship between internal factors and external stimuli: provided that there is some action-specific energy, the push of this and the pull of external stimuli will add up to give rise to the behaviour, rather than being multiplied together or related in some more complicated way. How does real behaviour match up to these four suggestions that Lorenz made? Let us consider each in turn.

## Accumulation of Energy

The nervous systems of animals have no stores of energy within them as Lorenz proposed, and this is obviously not a realistic aspect of the model. Nevertheless, animals might behave as if they did and, if so, we would expect them to become more likely to perform a particular action with the passage of time since they last did so. Do they do this? There is no doubt that certain aspects of behaviour do become more likely the longer the gap since they last appeared. Feeding and drinking are the most obvious of examples, as we all know from our personal experience. There are very good reasons for this: nutrients are used up by the body and water is lost constantly from it, so both must be replenished and the need to do so will rise with the interval since the last meal or the last drink.

But many of the other activities which animals perform are not concerned with regulating aspects of physiology, so there is no reason to think in advance that they might become more likely with time, and they are quite often found not to do so. Many behaviour patterns, such as singing in birds, mating in fish or exploration in rodents, could be used to illustrate this point. However, a particularly appropriate one to take is aggression. Although he did not mention his psychohydraulic model in his book *On Aggression,* Konrad Lorenz obviously had it in mind. He suggested that aggression is an innate drive which rises with time and must somehow be expended. His view of innate behaviour is that it is inevitable (a point taken up in the next chapter) and the only alternative is to sidetrack it into harmless channels: for human aggression he suggests that sport may play such a role, helping us to get rid otherwise destructive urges. Leaving aside the issue of whether sportsmen are less aggressive than other people as a result of their activities, there are a great many objections to these views. Some of them concern whether the urge to behave aggressively does accumulate in an inevitable fashion. One of the few cases where it seems to do so is where an animal is isolated, for example a mouse placed in a cage on its own. The longer it has been in isolation the more it will fight another mouse which is put in with it. But this could simply be that, when in company, it habituates to the other animals with it and it needs some days on its own to recover its tendency to fight them.

In fact, some fish fightless after isolation and need stimulation from others over a period before the urge to fight returns in this

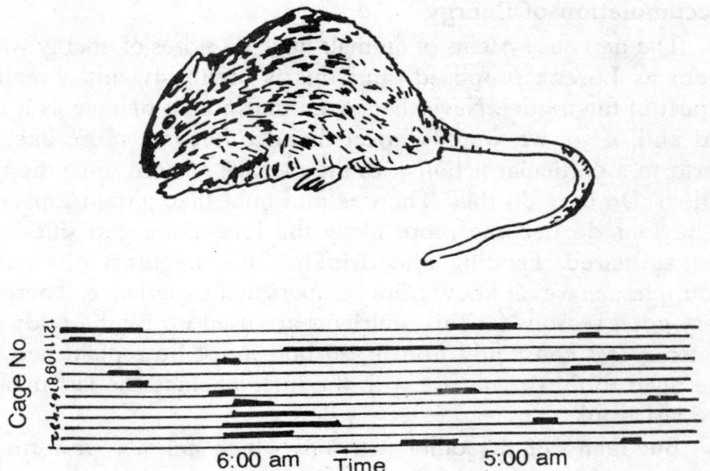


*Fig. 3.4. Traces from a pen recorder to show the feeding behaviour of 12 rats over a period of about 2 hours. The animals tend not to nibble at food the whole time, but pattern their feeding into meals, taking one of these at fairly regular intervals. Thus, the longer it is since its last meal, the more likely it is that a rat will eat.*

case contact with others probably leads to the gradual build up of a hormone which makes aggression more likely. Many factors affect aggressive behaviour, such as hormones and shortage of food on the inside, and presence of rivals and contested resources on the outside. Furthermore, aggression is itself a complex of different actions which may be quite differently caused though they are superficially similar. It is not an easy matter to decide the extent to which a hawk killing a mouse, a cornered subordinate rat defending itself against a dominant one, two fighting cocks locked in a struggle, and a mother duck defending her chicks from the unwelcome attentions of a gull, are actually motivated in a similar way.

Aggression serves many different functions within one species and its uses and the situations in which it appears may vary between species, so that the mechanisms underlying it are also likely to vary. Evolution matches behaviour marvelously well to an animal's particular environment and way of life. So it should be no surprise that the organisation of aggression various a lot between species just as feeding differs between the lion that kills and eats every few days and the wildebeest that must crop its plant food for most of its waking hours. Likewise, there is no reason why the urge to behave aggressively should mount with time like hunger or thirst, and it is rather surprising that Konrad Lorenz should have thought this given that he is a biologist with great respect for the power of evolution to adapt behaviour to an animal's exact requirements. A final aspect of the accumulating energy idea concerns the extent to which an animal needs to perform the action as opposed to needing to achieve its results. Must a hungry animal make the appropriate number of movements, as Lorenz's model would propose, or does it just need to have the right amount of food inside it? Does an aggressive animal have to amount of food inside it? Does an aggressive animal have to perform a certain amount of fighting, or will it cease to be aggressive when its rival is repelled? As already seen that animals tend to be aggressive when its rivalis repelled?

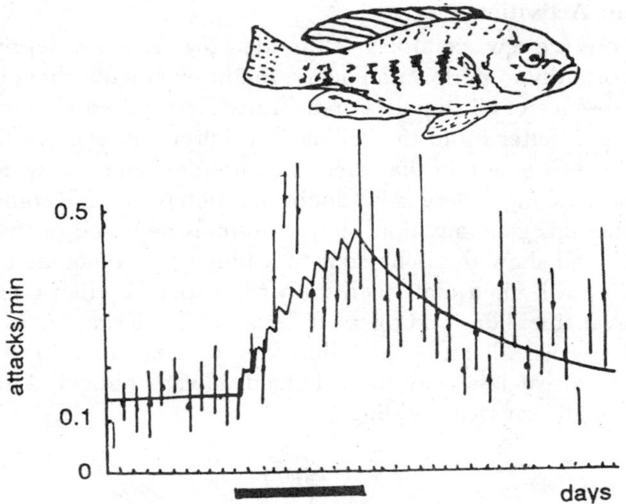


*Fig. 3.5. Males of the cichlid fish Haplochromis burtoni seldom attack small fish kept in their tank but, if a model of another large male is put in with them for 10 days (black bar), their attack rate slowly rises, then falls back again gradually after the model is removed. Thus, they do not attack most as soon as the model is presented, as would be expected had they accumulated action-specific energy in its absence, but instead need to be stimulated by a period of its presence before their attack rate-rises.*

As already seen that animals tend to regulate their behaviour as they go along in the light of its consequences: a process called feed-back. The dog does not rush at the place where the rabbit used to be but follows an arc so as to close upon it, changing its course as required. Unless its muscles are fatigued, it does not stop running before it reaches the rabbit, nor does it carry on after it has caught it. This is quite different from the action-specific energy idea, which proposed that the animal should perform the amount of behaviour for which it has accumulated energy regardless of its consequences. This would suggest that an eating animal should chew and swallow a set number of times even if an experimenter had surreptitiously raised the glucose in its blood or filled its stomach with food. In fact, animals do not generally do that: unlike the egg-rolling goose, they respond to feedback from their actions. They drink until receptors tell them that they have taken in enough water and they eat until they have had enough food, rather than showing a fixed amount of behaviour depending on the time since they last ate or drank.

**Vacuum Activities**

There are few examples of the idea that animals deprived of an opportunity to perform an action might eventually show it even in the absence of all stimuli. Lorenz himself described a pet starling that would flutter up to the ceiling to catch a non-existent fly, but it is hard to be certain that there was nothing there to which the bird was reacting. There is no doubt that behaviour can sometimes show stimulus generalisation, so that animals deprived of the usual stimulus will show the behaviour to a much less adequate one. In zoos, they will often mate with the wrong species when their own is not available: lions and tigers, for example, will mate to produce 'ligers'. As we all know, the hungrier one is the more one is prepared to eat less tasty food (indeed, if very hungry, I suspect even I might eat rice pudding!).

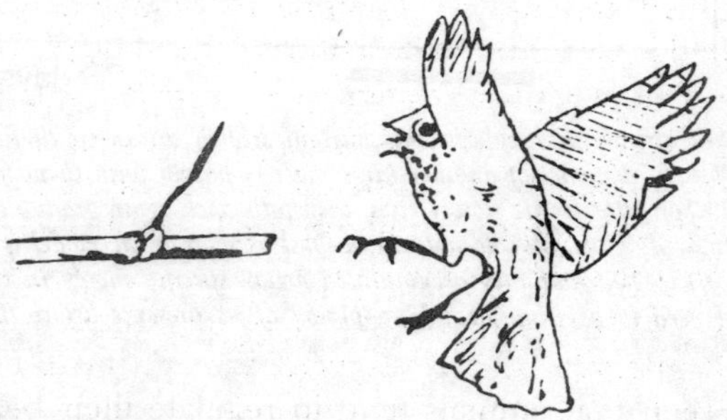

*Fig. 3.6. A European robin attacks the spot in mid-air where previously a stuffer rival had been placed.*

Thus, generalising to less adequate stimuli is a reality, but there is not so much evidence for behaviour being shown in the total absence of any appropriate stimulus, as the vacuum activity idea suggests. Perhaps the best is an account by the ornithologist *David Lack* of the behaviour of a European robin he had just finished testing with a stuffed bird of its own species, a potent releaser of aggression for these birds. As he removed the specimen and walked off, he chanced to look back and saw the robin fluttering around, singing and delivering pecks to the position in mid-air where its apparent rival had been. Here the appropriate stimulus had clearly been removed, but the animal was certainly not suffering from deprivation of the opportunity to behave aggressively.

## Deciding What to Do

Animals generally only do one thing at a time, yet they often have the need to perform several. For example, a caged zebra finch waking up after a 12 hours night must have a long list of priorities. These birds normally eat and drink about every half-hour, sing and fly around their cage for a period after this, and groom and rest till the next meal is due. Most of their grooming is brief, but every two hours or so they show a long bout which deals with all the areas of their body. At night they groom a little and rest a lot, but none of their other activities is performed during the hours of darkness.

Not surprisingly, then, they are rather busy in the first half-hour of the morning. Typically, they will stretch and ruffle their feathers when the lights come on, and then move over to their

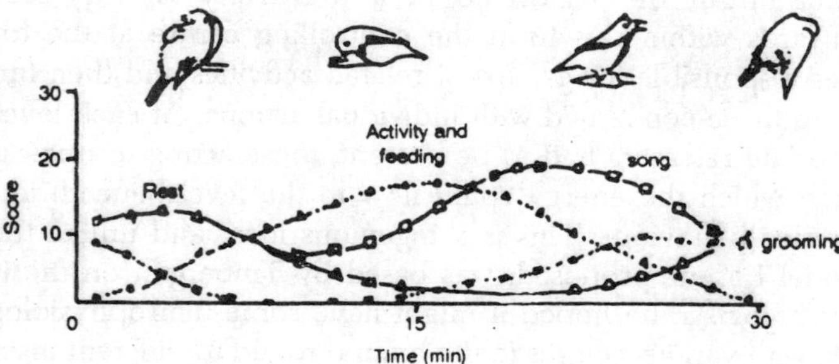


*Fig. 3.7. The short-term cycle of behaviour shown by a male zebra finch. The illustration is based on observation of 12 cycles shown by a single bird synchronised at 30 minutes long by a cycle of changing light intensity. In each cycle the bird moves from resting to an active period during which it feels, through a peak of singing to grooming and then back to resting once more.*

food for a long meal, perhaps followed by a drink. Some flying around and singing will follow this before they settled down for the first long grooming bout of the day. Despite their need to do several things, these birds show a well-organised sequence of behaviour. They do not rush around doing a little bit of one thing and a little bit of another, nor do they try to sing and groom at the same time. The sequence shown by different birds is similar and suggests that they all have the same priorities. The question posed in this section is how they decide between them, and it is a difficult question to which there is as yet no definite answer. What is clear is that the animal must make decisions and that to do so it must have some means of weighing against each other the internal and external factors relevant to different activities. The realisation that some sort of weighing up of opinions must go on within animals was one reason why psychologists postulated the existence of internal 'drives' which, like Lorenz's action–specific energy, were thought of as powering behaviour and were also seen as being measured against each other, the strongest being the one that was expressed. A similar idea was developed by early ethologists, though they tended to use the word 'instinct' rather than drive. Animals were thought of as possessing a small number of instincts, such as those for feeding, aggression and reproduction, and it was these that competed with each other.

The instincts themselves controlled a number of behaviour patterns: for example, the reproductive instinct led to various activities concerned with nest building, courtship and parental behaviour. Tinbergen developed a model suggesting that each instinct might be organised in a hierarchy Energy flowed downwards within this form the controlling centre at the top to centres responsible for groups of related activities and then further down to those concerned with individual actions. At each level the appropriate releasers had to be present, these acting to open gates through which the energy could flow to the level beneath and so on to give behaviour. This is a ingenious idea, and unlike the 'as if' model Lorenz proposed, was based by Tinbergen on the ideas of psychologists; he hoped it might have some neurophysiological reality, with various centres in the brain devoted to different instincts. Unfortunately, it turns out not to be that simple.

As mentioned before, the nervous system does not store and use up energy as these models suggest. Nor, unfortunately, it is

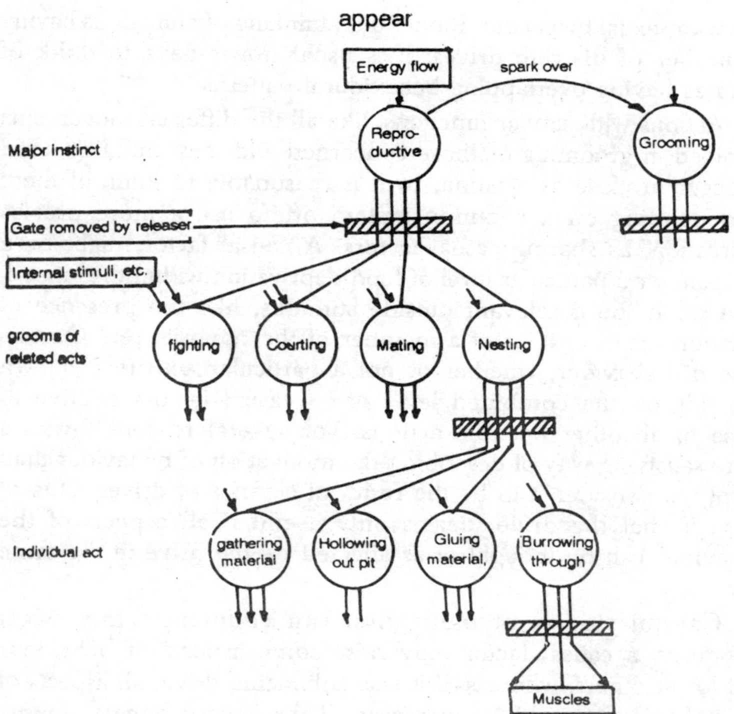


*Fig. 3.8. The essence of Tinbergen's hierarchical model was that a centre in the brain controlled each 'major instinct'. Energy from this flowed down to a series of lower centres when a gate was opened by the presence of appropriate releasers. In his original scheme, Tinbergen saw the hierarchy being extended right down to the level of muscle units. Though the hierarchies of different instincts were separate, lack of releasers for one could lead energy from it to 'spark-over' to another and so cause a displacement activity to occur. Thus, as in the example shown here, if reproductive activities were thwarted, grooming might appear.*

neatly compartmentalised into centres and pathways with clear and distinct behavioural systems, like those controlling feeding, drinking or sexual behaviour, can be though of as distinct, the factors affecting them overlap and they often influence each other. For example, the hormone oestrogen is an important internal factor making female rats receptive to the male, it also makes them very active so that they are more likely to come across a male, and it makes them less interested in food so that, when receptive, they spend less time eating and more looking for mates. This hormone, therefore, influences systems concerned with feeding, with activity and with sexual behaviour. Rather than thinking of animals as having a number of discrete drives, it is usual now-a-days to think of them as having overlapping behavioural patterns.

Actions with similar functions, like all the different movements involved in grooming or those concerned with nest building, tend to occur in close association, so it is reasonable to think of them as depending on a common system or, to use a more precise expression, as sharing casual factors. A causal factor might be a hormone or a particular level of food deprivation within the animal, or it might be a relevant outside stimulus, like the presence of irritation on its coat or of a member of the opposite sex. On this view of behaviour, whether or not a particular activity is shown depends on the combined level of its casual factors relative to those of all other possible actions. For several reasons this is a more satisfying way of describing the motivation of behaviour than simply to consider it to be the result of a series of drives. One of these is that the drive idea is only useful if all aspects of the behaviour can be thought of as affected by the drive in the same way.

Careful studies of motivation run counter to this. Most obviously, a causal factor may raise some aspects of behaviour and lower others, whereas, if it was enhancing drive, all aspects of the behaviour should be increased. Thus a very hungry animal may chew less and swallow more, and a bird building a nest actively may fly to and form with material at a great rate but spend less time than usual carefully selecting and gathering material or in weaving it into the structure. Another reason why viewing behaviour as dependent on a variety of internal and external causal factors is useful is that it helps to account for why behaviour patterns occur in association with one another to varying degrees.

A grooming animal will wash its face, scratch its flank and shake its body, and these very different actions tend to occur close together rather than in association with feeding or mating movements. The most likely reason for this is that grooming actions share causal factors with each other, such as activity in a particular part of the brain or irritation of the coat. Another reason why different behaviour patterns occur in association may be, quite simply, that one causes another. The fact that birds wipe their beaks after drinking almost certainly comes into this category: drinking makes the beak wet and so produces a stimulus which is only removed by wiping. Thus, as well as behaviour patterns sharing causal factors and thereby appearing close together, they can be linked because one acts as a causal factor for another. The way in which behaviour patterns may compete for expression has been more extensively studied in the case of feeding and drinking, because these are two acts the causal factors for which can be manipulated by depriving animals of food and water for different periods to time.

An animal deprived of water tends not to eat as much (doubtless dryness in the mouth makes it difficult to chew and swallow), so the two actions are not totally independent of each other. But just how hungry and thirsty a deprived animal can be established by seeing how much it cats and drinks when given food and water. We can then say it was 10 grams hungry and 1 gram thirsty if these are the amounts it consumes. The most detailed experiments on the relationship between feeding and drinking have been those on Barbary doves (known as ring doves in North America) by *David McFarland.* His research is a good example of how the use of techniques developed by psychologists has helped to illuminate ethological problems. The doves he studied were placed in the 'Skinner boxes' that psychologists often use to study learning, in which they could peck one key to receive water and another to receive food. He found that a dove deprived of both food and water would alternate eating and drinking until it had satisfied both requirements. Its food and water needs, and how these changed as it ate and drank, could be neatly illustrated in the form of 'state space' plots.

If the bird was more hungry than thirsty it would start by eating, and the amount that it ate could be traced by a line on the graph. After eating for a little it would then become more thirsty than hungry and would switch to drinking, the time on the graph moving so that the water deficit was shown as decreasing, while that for food remained the same. Finally, after several drinks and meals, the line would have zig-zagged down to zero on both axes, indicating that the animal was satiated. On this graph we can assume that every time the animal switches from one action to the other it is moving to that which has greater casual factors. We can draw a diagonal line which gives a rough impression of the boundary between feeding having priority and drinking doing so. All switches in the drinking segment are from feeding to drinking, and those in the eating one are in the other direction.

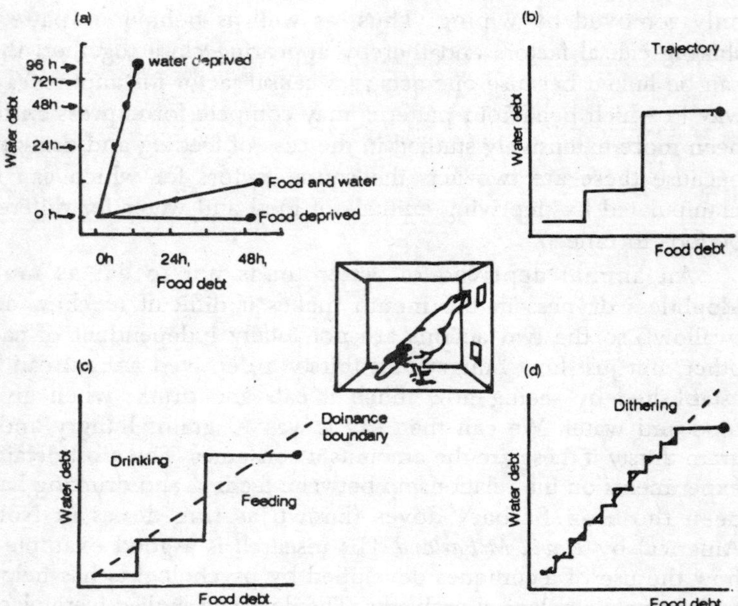


*Fig. 3.9. The way in which a deprived Barbary dove replenishes its reserves when placed in a Skinner box where it can peck keys to earn food and water. (a) The food and water debt of the bird can be represented by a point in a 'stage space'. Here three such points are shown for birds deprived of food alone, food and water, and water alone. Birds eat less when deprived of water, so the point for the last situation shows a small food debt as well, (b) A bird deprived of both food and water cats and drinks alternately, so tracing a trajectory in its state space which takes it down to the point where it has cancelled both debts. (c) The boundary drawn to separate areas where the animal starts to feed (food dominant) from those where it starts to drink (water dominant). The animal tends to feed until it is will into the water dominant area and vice versa, rather than dithering between the two, as shown in (d).*

A feature which may seem curious here is that the animal goes well over the line in each direction before it switches rather than doing so immediately. If it did the later, it would end up 'dithering': after an initial large meal or drink to take it over the boundary line, it would oscillate rapidly from one activity to the other, taking tiny amounts of each in turn until eventually both needs were satisfied. This would obviously be a badly organised way of doing things, but what makes sure that it does not happen? Several factors are probably involved, but one is likely to be especially important. This is that an animal that has just taken an item of food is likely to be looking at a food dish rather than at stimuli appropriate to any other action. As the sight of food is a casual factor for feeding, and that of water for drinking, this will mean that the animal is likely to continue with what it is doing rather than switching to another behaviour for which internal factors may will be high, but external ones are absent. In this simple example, the state space plot expresses the relationship between two activities, feeding and drinking, but a third dimension might be added so that the animal's tendency to groom was also mapped in the hope that one might be able to predict when it would take a break from feeding and drinking to clean itself.

Ultimately, one might hope to achieve mapping of all behaviour patterns onto each other in a 'multidimensional state space' but, as well as it being difficult to imagine such as complicated diagram, it is not easy to see what currency they could all be expressed in. There is certainly no such thing as 'grams' of grooming or of tendency to fight. So the problem of how the nervous system weights up different actions against each other remains as real as ever.

## The Physiological Basis of Motivation

Continued analysis of this behavioural type is essential, but is also important to examine motivation at the physiological level and try to link up behaviour with events in the nervous system. Some of the most promising bridges between neurophysiology and behaviour have developed from the work of physiological psychologists–most of them Americans–on motivational problems. *Grossman* provides a good introduction to the whole field of physiological psychology. We must now turn to consider some of this work and begin with some description of one area of the vertebrate brain which has proved to be of major importance in the control of motivation-the hypothalamus.

## The Hypothalamus

This relatively minute volume of brain tissue-in the human brain it is smaller than the last joint of the little finger-is of primary importance in a whole host of reactions. There is an excellent general account of its physiology in *Walsh* who says the hypothalamus,'... This small centre plays a dominant role in determining the use that is made of the resources of the body.... It is difficult, indeed, to think of any function of the body that it is not dependent, directly or indirectly, upon the hypothalamus'. A little neuro-anatomy is needed at this point; the reader is referred to a clear, concise account in *Romer.* The brain of all vertebrates is constructed on the same basic plan, and at an early stage in embryology consists of three swellings at the anterior end of the spinal cord. These are called the prosencephalon, mesencephalon and rhombencephalon or more simply the fore, mid-and hind-brain. Primitively these swellings arose to cope with the increased amount of sensory information which flowed into the central nervous system from the sense organs of the head.

The fore-brain originally dealt with olfaction, the mid-brain vision and the hind-brain balance and hearing. In most living vertebrates their original functions have become greatly extended and complicated, but still the appropriate sensory data are led first to these regions, even if subsequently they are passed on elsewhere. The fore-brain is easily sub-divided into two portions. The anterior portion has arising from its roof the cerebral hemispheres, which primitively were olfactory areas but have now come to dominate the whole nervous system in mammals. The posterior portion of the fore-brain–called the *diencephalon*–has on its dorsal surface the pineal organ. This was once associated with a light receptor or pineal eye which can still be seen in some living reptiles.

The side walls of the diencephalon are thick and form the thalamus, an important 'staging place' in the brain where fibre tracts link up with one another in numerous 'nuclei' or clusters of neutron cell bodies. On the floor of the diencephalon, below the thalamus as its name implies, it the hypothalamus. There are nuclei in the hypothalamus, but they are not so well defined as in the

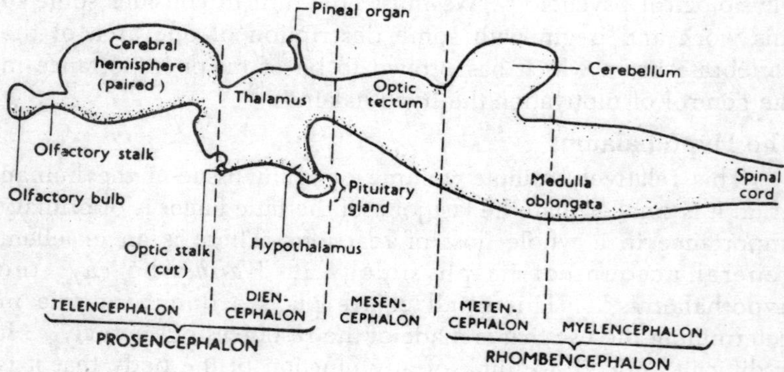


*Fig. 3.10. The basic division of the vertebrate brain. The brains of all vertebrates pass through a stage rather like this during development, but in mammals and birds in particular, the adult brain is dominated by the enormous growth of the cerebral hemispheres and cerebellum.*

thalamus above. However, there are several well-marked fibre tracts entering and leaving, and these put the hypothalamus into connection with the cerebral hemispheres and also with more posterior parts of the brain.

From the behaviour point of view, one of the most significant structural features of the hypothalamus is its intimate connection with the pituitary gland. This endocrine gland, which controls the whole hormonal system of the body, develops from the fusion of a down-growth of the embryonic hypothalamus with an upgrowth from the roof of the embryonic mouth cavity. The pituitary stalk, which joins it to the hypothalamus, contains both nerves and blood vessels. The hypothalamus itself has a very rich blood supply and some of its cells are even penetrated by capillaries. From its connections with other parts of the brain, its rich blood supply and its links with the pituitary gland, the hypothalamus is well adapted both to measure changes in the metabolism of the body and to set in motion activities which will rectify them.

It, in other words, well suited to serve as part of a homeostatic control system and there is plenty of physiological evidence that it does so. For example, the control of body temperature is one of the most delicate homeostatic systems in a mammal or a bird. There are areas of the hypothalamus which are highly sensitive to changes in the temperature of the blood. If these areas are heated artificially by implanted wires, the animal states sweating and painting. Sweating is controlled peripherally by the autonomic nervous system and the hypothalamus can initiate its activity. The reverse effect is produced when the temperature-sensitive areas cooled; now the animal shivers–another autonomic response (*Walsh*). All this many appear to be pure physiology, having little to do with the study of behaviour. But homeostasis is one of those topics were the boundaries between traditional fields of study break down. It is not helpful to distinguish between physiology and behaviour in the matter of temperature control. If a rat is briefly cooled in the manner described, shivering is started to generate some heat. We might classify this as a reflex activity and consign it to the realms of physiology. However, if the rat is cooled for a longer period shivering alone in inadequate and, if given the material, the rat begins to build a nest or to enlarge the one is already has, in order to insulate it self. The reflex response is not supported by a complex behavioural one. Both are initiated by the hypothalamus and both form part of the rat's homeostatic system, although nest building will involve a more elaborate neural mechanism than does shivering.

## The Hypothalamus and Motivation

It is from studies of the role of the hypothalamus that some of the most important links between brain and behaviour have developed. Modern physiological techniques allow parts of the brain to be explored with electrodes, controlled injection of chemicals or by the destruction of very small selected areas. As a typical example of the way such studies reveal the underlying physiology of motivation we may consider the role of the hypothalamus in thirst.

There are cells in its lateral areas, which respond to increased concentration of the circulating body fluids. They can set in action two compensating systems. The first, acting via the links with the pituitary gland, causes the secretion of *Antidiuretic Hormone* (ADH), which increases the resorption of water by the kidneys. The second system causes the animal to seek out and drink water. If the link between the hypothalamus and the posterior lobe of the pituitary gland is damage, ADH may never be secreted. In such a situation the kidneys continue to excrete copious quantities of urine and to compensate the animal drinks large quantities of water–a condition known as *diabetes insipidus.* Normally, the amount of water taken in is exactly adjusted to the animal's needs, by if the lateral hypothalamic detector area is artificially stimulated either electrically or by injecting hypertonic saline, drinking is greatly increased.

*Anderson* prepared a number of goats with fine hollow needless penetrating into the hypothalamus. He allowed them to drink as much water as they would take and then injected concentrated salt solution. There was little effect unless the needle tip was in the lateral hypothalamus. In this region the salt cause the goats to drink frantically within a minute or two of injection. These animals, previously satiated with water, would drink salt and bitter solutions, which they would not normally touch even if extremely thirsty. This condition is not comparable to diabetes insipidus, because the goats are not compensating for water lost through the kidneys. They are drinking water which is surplus to their physiological needs. *Andersson* could produce the same effect if the stimulated the lateral hypothalamus electrically and such drinking behaviour is often described as 'stimulus-bound', because it continues only whilst the stimulus (saline or electrical) is actually being applied.

Stimulus-bound drinking has also been produced in the rat and the importance of the lateral hypothalamus for drinking is further emphasized by the behaviour of rats with damage to this part of the brain. Such animals may stop drinking altogether. They will not drink even though in the last stages of dehydration and will eventually die in the presence of water, unless this is given artificially through a tube into the stomach. It is fascinating that a rat with a lateral hypothalamic lesion is not just disinterested in water, it becomes actively averse to it. If water is placed in its mouth it will not swallow but allows the water to run out, making tongue and lip movements identical to those made by a normal rat towards an intensely bitter liquid. These effects of stimulation and ablation might lead us to conclude that the lateral hypothalamus is responsible for the state we observe behaviourally as a tendency to drink. How far is this conclusion justified?

Before discussing this question we can add comparable results which also implicate the hypothalamus in the control of feeding. Mammals and birds normally keep their weight very constant and adjust the amount they eat accordingly. If rats are given a super-rich diet they eat less; if their food is mixed with non-nutritive cellulose they eat more. We have already mentioned that rats with damage to central areas of their hypothalamus (the ventromedial nucleus, in fact) lose this sensitive control of their eating.

The food intake of a rat whole ventromedial nucleus has been ablated compared with that of a sham-operated animal. After a few days of post-operative depression of food intake, the brain-damaged rat begins to eat huge amounts of food–at least four times the normal amount. This so called 'dynamic phase' of hyperphagia lasts for 3 weeks or so. Beyond this, food intake slowly declines and eventually settles down with the animal eating about double the normal amount. Needless to say hyperphagic rats become grotesquely fat and are very inactive. Electrical stimulation of the ventromedial nucleus of normal rats depresses their feeding, so that this region of the hypothalamus has often been called a 'satiety centre'; it measures when the animal has eaten enough and inhibits further eating. (The term 'centre, here means a group of nerve cells with a common organizing function.

We may note that hyperphagic rats do not-lose all control of their food intake. That eventually they are eating considerably less than they did during the early dynamic phase. Station is not

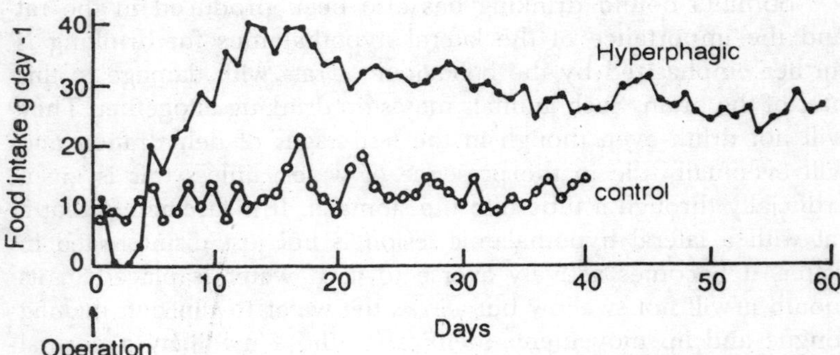


*Fig. 3.11. The daily intake of normal rats and those with bi-lateral lesions in the ventromedial nucleus of the hypothalamus.*

abolished with the ablation of the ventromedical nucleus, but its threshold is greatly raised. A centre complementary in function to the ventromedial nucleus and which promotes feeding is located in the lateral hypothalamus closely associated with the drinking area. In rats, experiments similar to those described for thirst have shown that this 'feeding center' does initiate–the appetitive behaviour of eating: stimulus-bound feeding can be elicited by implanted electrodes and damage to the area can lead to starvation in the presence of food.

The feeding and drinking centers are closely linked anatomically and they interact with each other in a complex fashion. For details of the behaviour associated with this interaction the reader is referred to the admirably clear account by *Teitelbaum* and *Epstein*. How far are we observing the operation of the 'normal' drinking and feeding systems during stimulus-bound behaviour–remembering that this term does not refer to 'normal' stimuli but only to artificial stimuli applied chemically or electrically within the brain itself? It would be important to see whether an animal shows normal appetitive behaviour when being stimulated, because it might be argued that their feeding or drinking was merely a reflex response to the stimulation of neural pathways controlling the motor patterns concerned. It is possible to get fairly well co-ordinated lip and tongue movements by stimulating other areas of the brain–parts of the motor cortex of the cerebral hemispheres, for example. Some of Anderson's observations appear to rule out this explanation because, when stimulated, his goats showed all he signs of normal thirst. They walked over to the corner of their pen and searched around for the water bowl, i.e., they showed normal appetitive behaviour and not just 'forced drinking'.

Again with feeding in rats, coons have shown that stimulus bound feeders will learn a new response (bar pressing) in order to acquire food, and that this response is transferred and used when subsequently the same animals are made normally hungry. Such results are very convincing, but some doubts still remain and have been emphasized by *Valenstein* and others. There are a number of experiments which have shown that stimulus-bound feeders and drinkers are much more easily dissuaded from their goal than normally motivated animals. At least this is the case in rats, where slight adulteration of food or water with quinine is usually enough to stop them responding. (We have already mentioned how easily hyperphagic rats are dissuaded by quinine).

Further, the rate at which rats drink by lapping from a tube is normally very constant. Normal water deprivation simple leads to lengthening their bouts of drinking but does not affect their lapping speed. However, *White* report that stimulus-bound drinkers do change their rate of lapping with the intensity of electrical stimulation, which certainly suggests that the motor-organization centers for drinking are being affected. We shall need more facts to be able to resolve this question and, even from the evidence we have at present, we must not expect the same details to apply to all vertebrates or even to all mammals. Nevertheless, there can be little doubt that the basic function of the hypothalamus as a detector of physiological imbalance remains constant. Its cells will meter the temperature, food, water and hormonal concentrations of the bloodstream. As a result of any particular imbalance, the detector sites will initiate both physiological and behavioural action, as we have outlined earlier in this section. The behavioural problems then become more obscure, because whilst we may accept that the hypothalamus is essential for the initiation of motivational states, we cannot claim that it alone is essential for their control. Clearly this control involves many others areas of the brain also.

Grossman reviews the extensive evidence showing that other parts of the forebrain are involved in feeding, drinking sexual and aggressive behavioural effects. It has been found that damage to most areas of the cerebral hemispheres does not affect an animal's motivation in a specific way, but there are some significant exceptions. Many of theses deprivation result in behaviour which is designed to restore such deficits and we noted that together with behavioural responses there may also be hormonal ones; thirst leads to the release of antidiuretic hormone from the posterior pituitary. We can now consider the interaction between hormones and behaviour in more detail. This topic deserves closer attention because in many ways the endocrine organs, which produce the hormones, and the nervous system share the common functions of communication and co-ordination both within the animal and between it and the outside world.

The hormones form a chemical message system, which is probably as old as the nervous system. Indeed one may have developed from the other in part. Throughout the animal kingdom we find neurosecretory cells in the nervous system. These are modified neurons which can pass special chemicals down their axons and into the bloodstream. Often these cells are clustered together to form glands, such as the corpus cardiacum of insects, which have close connections with both nervous system an blood stream. The vertebrate pituitary gland develops from the fusion of neural and epithelial tissue and remains closely connected to the hypothalamus. The pituitary regulates by its various secretions all the other endocrine glands and is in turn regulated by the nervous system. This system of control enables environmental changed picked up by the nervous system to be matched by an appropriate hormonal response. The most familiar example is the control of breeding seasons in mammals or birds.

Changing day length, perceived by the eyes, results in changed activity in the hypothalamus which stimulates the pituitary. This in turn secretes hormones, which start the various growth changes in the body associated with the onset of breeding condition. Throughout all their interactions the functions of the two communication systems–endocrine and nervous–remain essentially complementary. The nervous system can only pass information by trains of nerve impulses. Its state can change very rapidly, but it is clearly less suited to transmit a steady unchanging message for a long period; it operates on a time scale from milliseconds to minutes.

The endocrine system cannot respond so rapidly, but its cells cam maintain a prolonged steady secretion into the bloodstream lasting for months if necessary. Moreover hormones can reach every cell in the body via the bloodstream, whereas the nervous system generally controls only the muscles. Circulating hormones are commonly regarded as prime motivating factors in animal behaviour and ethologists have often assumed the other act directly on brain control centers to increase drive. Certainly, there are dramatic examples of hormones, apparently 'forcing' behaviour from an animal even in the most inappropriate circumstances.

For instance, *Blum* and *Fiedler* describe how injections of the pituitary hormone prolactin will cause isolated males of the fish *Crenilabrus ocellatus* to perform the parental fanning movement. This is similar in form and function to the stickleback's fanning. Injected *Crenilabrus* males will fan in a bare tank devoid of all the normal external stimuli for funning, $CO_2$ in the water, fertilized fanning is dependent on the dose of prolactin. Examples of this type are convincing evidence for the central motivating role of hormones but they can also affect behaviour in other ways. Before discussing examples of hormone action it is necessary to give a brief outline of those aspects of the vertebrate endocrine system, which are most important for behaviour. Fuller account are given by *Gorbman* and *Bern* and by *Austin* and *Short.*

## The Pituitary Gland

The pituitary gland secretes several hormones which affect the output of other endocrine organs and in this way the pituitary effectively controls the whole endocrine system. The hormones we shall be most concerned with are the *gonadotrophins* which act upon the gonads and promote both the growth of germ cells and the tissues of the gonads which secrete the sex hormones. There are two main gonadotrophins–*Follicle Stimulating Hormone* (FSH) and *Luteinizing Hormone* (LH)–both were named after their action on the female gonads or ovaries but they are secreted by, and have similar functions in male's. In females both are necessary for the growth of eggs and for their release into the oviduct ready for fertilization. A third pituitary hormone important for behaviour is *prolactin*, also known as *lactogenic hormone* or *luteotrophic hormone* (LTH), which has a variety of physiological effects. We know that it is secreted by various classes of vertebrates but it often has completely different functions and 'target organs' (those parts of the body whose growth of functioning is affected by the hormone).

Interestingly most of its effects are on ' parental behaviour' in the broadest sense. As just mentioned, prolactin stimulates fanning of the eggs by male sticklebacks and some other fish; it promotes broodiness in chickens (although not all birds), the secretion of crop milk in pigeons and the growth of the mammary glands and milk secretion in mammals. The name luteotrophic hormone refers to another important function of prolactin in mammals, for it is required for the maintenance of corpora lutea in the ovary and stimulates them to produce progesterone.

**The Gonads–Ovary and Testis**

It has been known for centuries that castration has profound effects on the behaviour and body form of vertebrates. This is in contrast to many invertebrate groups in which castration has little or no outward effect. Only in the vertebrates are the gonads important endocrine organs where, under stimulation from pituitary FSH and LH, they produce the sex hormones from special secretory cells. The female hormones are collectively called *oestrogens* and the male hormones, *androgens.* All these hormones are steroids which are closely related in chemical structure, and although different vertebrate groups secrete slightly different steroids, those from one group are usually quite effective in another. The commonest androgen secreted by mammals is called *testosterone.* The sex hormones are responsible for the development of the secondary sexual characters and for the growth of the reproductive systems in preparation for the shedding of eggs and sperm.

Usually there are permanent differences between the body form of male and female, which are maintained throughout adult life. These are often augmented by the seasonal growth of secondary sexual characters under the influence of increased sex hormones. Stags can be distinguished from hinds all the year round but in addition they show seasonal growth of antlers. Finally, we must mentions another steroid hormone produced by the ovary. After a mammalian egg has been shed, its empty follicle enlarges and forms a prominent yellowish structure on the ovary surface–the corpus luteum. This begins to secrete *progesterone* under whose influence the lining of the uterus is prepared for receiving the egg after fertilization and development into a blastocyst.

Progesterone also inhibits the contraction of the uterine muscles, which must be avoided if pregnancy is to continue. It is justly called 'the hormone of pregnancy' but it is not an exclusively mammalian hormone. Structures like the corpora lutea from when egg are shed in fish, amphibians and reptiles but we know little about the presence or action of progesterone in these classes. Birds have progesterone which is almost certainly secreted by the ovary although their corpora lutea are not conspicuous: male birds of several species are also known to produce progesterone, probably in the testis.

## HORMONES AND BEHAVIOUR

In the context of this chapter it is possible to give only the most superficial account of what is known of the role of hormones in motivation. Of all the internal factors the influence behaviour, hormones are the most fully understood. There are two reason for this. First, hormone levels in the body can be precisely measured and related to behavioural changes. Secondly, and more significantly, the endocrine organs that secrete hormones can be removed with little damage to the animal. It is than possible to examine the role of a specific hormone by injecting known quantities of it, in effect mimicking the secretion of the removed organ, for example, the role of testosterone in the behaviour of male animals is commonly studies by injecting it into castrated individuals. Hormones are general factors in motivation in two sense. First, they typically influence several aspects of behaviour and may be involved in the causation of a number of different activities. An obvious example it testosterone, which influences not only sexual behaviour but also aggression. It appears that testosterone may have other effects. In chicks and mice, it increases the persistence with which individuals search for a particular type of food (*Andrew* 1976). In courtship also, some of the effects of testosterone may be because it increase the male's persistence. The second sense in which hormones have general effects is that they influence behaviour over a relatively long time-scale.

Reproductive hormones are generally necessary for the expression of sexual behaviour, but levels of these hormones do not change from moment to moment, so that changes in sexual behaviour cannot be attributed to concurrent changes in hormone levels but must be explained in terms of time-scaled over which most hormones act on behaviour may be an animal's lifetime, as in sexual maturation, one or more years, as in animals that breed more or less annually, or a matter of days or weeks, as in the control of reproductive cycles in mammals. The 'fight or flight' hormone adrenaline, which brings about the effects one feels which suddenly angered, is unusual in having effects on behaviour over a matter of minutes or even second, but future research may well

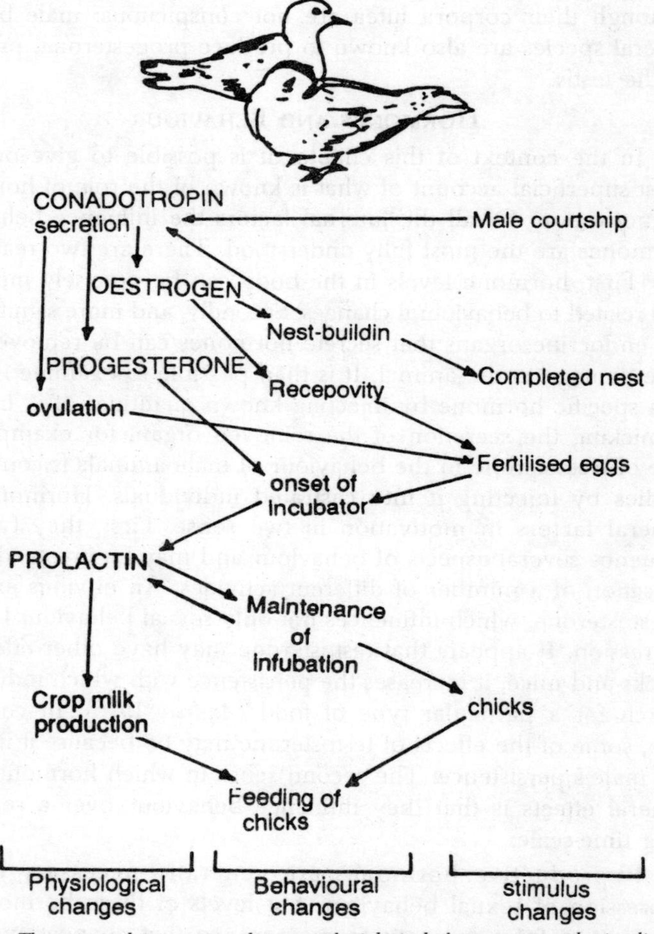


*Fig. 3.12. The principal interactions between physiological changes, external stimuli and behaviour involved in the reproductive behaviour of female ring doves. Hormones are shown in capitals.*

reveal other such short-term influences of hormones on behaviour. Although the effects of hormones on motivation are obviously internal, it is important to stress that they do not work in isolation but interact in complex ways with causal factors from the outside.

Hormone secretion is commonly itself dependent on external stimuli; for instance, in male birds testosterone secretion starts in spring in response to increase daylength. Also a hormone cannot usually elicit behaviour on its own without appropriate external stimuli, such as those provided by a mate. One of the most elegant features of the relationship between hormones and behaviour is the way in which the hormones appropriate to an aspect of behaviour have often been found to be secreted in advance of the situation in which that behaviour appears. This is a good example of feed forward mechanism. One of the most complete analyses of the interaction between hormones, external stimuli and behaviour is that which has been carried out on the reproductive behaviour of doves.

Originally started by *D.S. Lehrman*, this work has been carried on by many others *Silver* (1978) and by *Cheng* (1979). If a pair of Barbary doves (*Streptopelia risoria*) are housed together with some nest material and a bowl in which to build a nest, they will go through a week-long courtship period, towards the end of which they mate and build a nest. Two eggs are then laid and, until these hatch about two weeks later, male and female share in incubating them. By the time hatching occurs, the crops of both parents have developed the capacity to secrete a milky fluid on which the nestling called squabs, are fed. The behaviour of a female dove is influenced by a number of physiological factors, including at least four different hormones, and by external stimuli from her mate, the nest, eggs, and squabs. A feature of this system is that at each stage the female is stimulated to secrete the hormones that prepare her for the next stage.

Male courtship stimulates the secretion of oestrogen and progesterone, which cause her to become sexually receptive and which are necessary both for incubation to be maintained and for the secretion of crop milk. That prolactin secretion is stimulated by and stimulates incubation indicates positive feedback between the behaviour and a casual factor that underlies it. This mechanism ensures not only that incubation is maintained but also that when it ends and the eggs hatch, the birds are secreting crop milk and are thus ready to feed the young immediately. The reproductive behaviour of male doves appears to be controlled in a rather different way from that of females, although prolactin has similar effects on both sexes. Whereas it is clear the testosterone, which shows a marked rise after pair formation, is involved. Many of the changes in a male's behaviour are primarily responses to the female's activities, not to his own hormonal state. For example, he begins nest building in response to the female starting to spend long periods at the nest site, not to any specific endocrine evens (*Erickson* and *Marines-Vargas* 1975).

It appears that testosterone is involved in the causation of a wide range of male activities, including several different courtship displays, nest building and incubation, but that their precise coordination is determined by external cues from the female. The nest and its contents, perhaps what this example illustrates most clearly is that it is a mistake to think of causal factors as either specific or general because some of them are more limited to just one of them are more limited in their effects than others. Testosterone does not affect all activities performed by male doves, not are its effects limited to just one of them, but it does influence a variety of activities of varying degrees.

**Displacement Activities**

The idea that animals sometimes show displacement activities was a final proposal made by ethologists which deserves discussions. Watching a pair of courting birds, the observer might be surprised to see one of them break off to preen briefly and hurriedly before carrying on with its courtship. Similarly, a fish in the middle of threatening a rival may suddenly swim down and start digging in the sand as if switching to nest-building, or a cock might interrupt fighting to take a few pecks of food. Such actions seemed irrelevant to those who saw them and they often shared other features such as incompleteness or a hurried and frantic appearance, so they came to be grouped together and labelled as displacement activities. It was suggested that they arose when the animal was unable to carry on with more relevant behaviour because it was in a conflict as to what to do or was thwarted from achieving its aims. Thus, the fighting gull might be in a conflict between attack and escape and so unable to do either; a courting male fish might have his sexual approaches thwarted by an unreceptive female and so, again, he unable to show the behaviour most relevant to the moment. In line with his hierarchical model of motivation.

Tinbergen suggested that the thwarted energy from the reproductive instinct 'sparked over' and motivated behaviour down a different channel, such as that concerned with grooming. Ideas on displacement activities have changed enormously since they were thought to be due to sparking over. An important finding was that the causal factors which affect the behaviour in its normal context also affected it when it appeared as a displacement activity. A

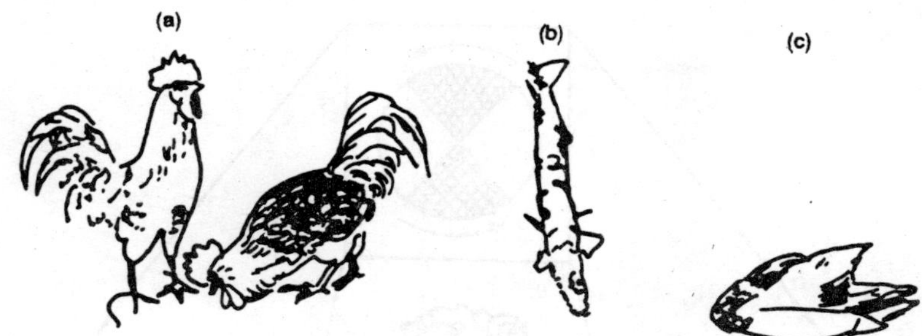


*Fig. 3.13. Three classic examples of displacement activities: ground pecking in a fighting cock (a); sand digging in a three-spined stickleback confronted by an intruder (b); wing preening in a courting mallard drake (c).*

courting male stickleback might break off to fan at the nest, a behaviour which serves to aerate his eggs but which is irrelevant when there are no eggs there yet. Nevertheless, in its courtship context as well as when it appears later, the fanning is enhanced by carbon-dioxide in the water.

In the same way a bird which grooms when in conflict shows more of this grooming when its feathers are wet. Results such as these show that the behaviour patters labelled as displacement activities are motivated in the usual way: their irrelevance lies only in their context. The behaviour patterns labelled as displacement activities are undoubtedly a rag-bag of different actions, differently caused, the main thing they have in common being that a watching ecologist thought they were out of place. Conflict or thwarting may certainly be one reason shy they arise: the animal, unable to carry out actions which are highest in priority, moves instead to ones which are next in line. This process is called disinhibition, and it is probably an important reason why displacement activities appear.

It is perhaps not surprising that very often the displacement activities that have been described are grooming with it so that, unlike food or water, they are always there on the outside of its body. Thus, if disinhibition occurs, grooming is very likely to follow, however, another reason why grooming often occurs out of context may be much simpler: the vigorous actions of fighting or courtship that the animal has been showing may lead its fur or features to be dishevelled and so actively enhance the stimuli for grooming till it breaks off for a quick preen to get rid of the itch. In the days when ethologists thought of behaviour as being motivated by

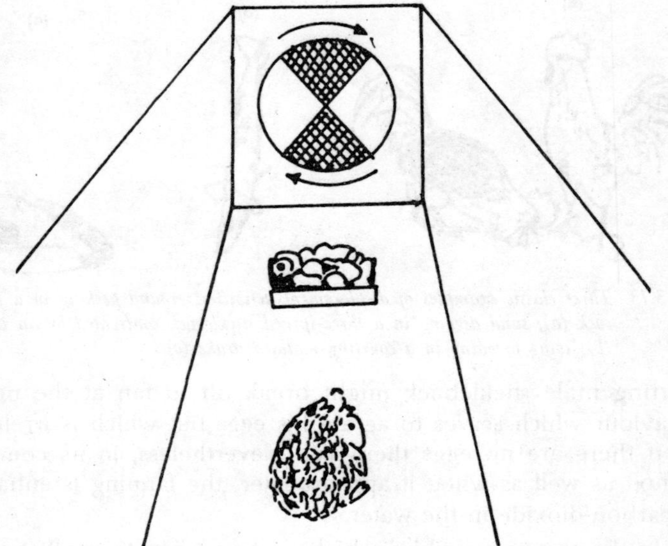

*Fig. 3.14. A hungry guinea pig placed in a conflict by a rotating card near its food dish stops some distance from its goal and shows various behaviour patterns such as grooming.*

instincts or drives, it was these concepts that were thought to be in conflict with each other within the animal, or to be thwarted by a barrier to its behaviour on the outside. Thus conflict between the drives of aggression and of escape, or thwarting of the sex drive, were commonly referred to. However, these ideas were very vague; drives themselves were hypothetical, so the conflict between them was doubly so.

The theory was also virtually untestable: if everything was due either to its own drive or to a conflict between other drives, it was hard to think of any observation or experiment that could disprove it. Thus, as drive theories have fallen from favour, so has the idea of conflict between drives. But this does not mean that conflict or thwarting is not a real phenomenon and an important cause of switches in behaviour. Thwarting can easily be arranged by placing a perspex screen between a hungry animal and its accustomed food dish. Similarly, an approach-avoidance conflict can be set up by placing a frightening object beside a food dish. In such circumstances a whole array of behaviour patterns may bee disinhibited because the animal is unable to perform those that are its top priority.

4

# Learning and Habituation

Learning–the modification of behaviour in response to experiences–is one of the most characteristic attributes of animals. The most important feature of learning is that it is adaptive. The animal, having learned, responds in ways that improve its survival and reproductive success. Different animals, even of the same species, learn different things if they are exposed to different environments. Learning is demonstrated when animals in two groups, one given an experience denied to the other, develop different behaviour patterns. Sometimes these experiments reveal that experiences do modify behaviour, but at other times they show that experience is unnecessary. For example, most young birds begin "practice-flying" before they leave their nest. They stand tall and flap their wings vigorously as if they were testing them for the first take off. But experiments have shown that birds can develop the ability to fly purely by maturation.

In one experiment, pigeons were reared in narrow tubes that prevented them from moving their wings. They could not undertake any form of practice flights. A control group of birds were allowed to practice each day. When the controls had "learned" to fly, the experimental birds were released from their tubes. Surprisingly, they flew just as well as the control birds. According to *Hilgard* (1956) learning occurs when the probability of certain behaviour patterns in specific stimulus situation has been changed as a result of previous encounters or other similar stimulus situations. *Hinde* (1970) has defined the learning as changes which can not be understood in terms of maturational growth process in the nervous system fatigue or sensory adaptations. According to *Wallace* (1973) learning is the adaptive changes in behaviours that result from the individual experience and probably associated with physical changes in the central nervous system. It may function with innate pattern or but may ultimately be separable from these patterns.

## Learning, Adaptation and Natural Selection

It is also reasonable to suppose that behaviour and psychological processes such as learning and memory play a role in adaptation and may also be selected for. One way to visualize the role that behaviour may have played in evolution is to consider again some of the problems that an individual must solve in order to reproduce. Consider, for example, the necessity of food. If food tends to be unevenly distributed in the environment, it would be to the animal's advantage to be able to learn and remember routes to the best food sources, cues that predict the availability of food, the route back to the nest (if there is one), signs that predators might be in the vicinity of the food source and, if so routes to alternative food source. Also important would be the ability to learn and remember the distinguishing characteristics of toxic food substances. There is substantial evidence that learning is, in fact, important in food selection and foraging behaviour, and we will consider this in some detail in several places in this text.

For now it is important only to imagine that animals particularly efficient at these tasks would be more likely to produce offspring and hence pass on the ability to learn. Is learning in regard to food-related behaviour important in human culture? Just think of the substantial time allocated to such things as learning how to cook, recipe preparation, comparative food shopping, agriculture (requiring entire college curricula), the technology of fishing, and the dairy industry. Anti-predator behaviour is another problem faced by animals. Many animal species have evolved elaborate structural defenses such as camouflage colorations and morphology, and mimicry of toxic or dangerous animals. Some have developed evasive running behaviours; other have group defenses such as schooling (fish), herding (grazing animals), or mobbing (birds).

Learning also evolved to play a role in the safety of animals. For example, mobbing of a predator by birds not only serves to drive off the predator but also to acquaint the young with the nature of typical predators. For example, one species of monkey has different alarms calls for different types of predators and these calls must be learned by the young. It is interesting that the accumulated knowledge of human culture and technology is used for anti-predator behaviours in ways that mimic the structural and behavioural adaptations of other animals. For example, in warfare (humans are essentially the only predators that humans have to fear) protective colouration is used. Ski troops wear white, jungle troops wear green, desert troops were khaki. There is even a "stealth" aircraft under development that is "invisible" to radar. Speed and evasive behaviours are used (in aircraft and ships, in particular). Repellent devices are used (weapons). All of these strategies employed by humans are based on learning in the sense of accumulated technology. It is not necessary to elaborate any further on how learning may have evolved (been selected for) in other survival problems that all animals face problems such as care of the young, habitat selection social interactions with conspecifics, etc.

The principal point is that survival problems can be solved in several ways by structural changes that adapt the animal to its environment; by innate behaviours based on complex reflexes or by behavioural changes based on the ability to learn. There is an important difference among these modes of adaptation. Adaptation resulting from selection of the ability to learn and behave flexibly provides an enormous advantage in that it contains the basis of further adaptations in the life-span of a single individual. An organism that has adapted solely on the basis of structural modification or innate behaviours faces difficulty if the environment changes; an organism that has adapted on the basis of learning may be able to learn new behaviours that will adapt it to the new environment. If this learning takes place, then such an individual will pass on its genes to future generations whereas the organism that has adapted solely on the basis of morphology or fixed behavioural mechanisms would be less likely to do so. Thus, the ability to learn would have been selected as an adaptive mechanism. Perhaps the flexibility of learning can be emphasized by describing some examples of complex behaviour that are surprisingly inflexible.

A species of wasp provides food for its offspring by paralyzing its insect prey and depositing it in an underground nest for the larvae to feed upon. In doing, this the wasps goes through a regular sequence of behaviour. When it first returns to the nest with its paralyzed quarry, it places the insect near the entrance and then goes down into the nest itself. The wasp soon returns, brings the insect down into the nest, and returns to the surface to close up the entrance. However, an interesting behaviour occurs if, while the wasp is on its initial inspection tour" of the nest, its prey is moved several inches away from where it was left at the edge of the nest. When the wasp now returns to the surface, it moves the prey back to the entrance and then goes down into the nest again–starting the sequence over. This apparently can go on and infinitum as long as the prey is moved each time the wasp enters the nest. It is as if the wasp were caught in some program "loop" that must be completed before it can go to the next stage of its provisioning sequence. The wasp, in this instance, demonstrates remarkably inflexible behaviour. Another example of inflexibility is the behaviour of the wasps Ammophila pubescens. This wasp may maintain many burrows, with larvae at different stages of development in each.

The wasp provides its young with food, giving each larva an amount appropriate for its size. Before setting off on the day's foraging the wasp visits all the burrows and them, eventually, returns with the proper amount of food for each burrow. If however, the larvae are switched after it has made its first visit, the wasp does not adjust its behaviour to the new occupants of the burrows. Instead, the amount of food placed in each burrow corresponds to the size of the offspring that has been there on its first visit–it does not correspond to the size of the current occupant. Again, the behaviour of the wasp is complicated and is a performance that requires a considerable memory. But the behaviour is also stereotyped and resistant to change based on changed conditions. In considering the evolution of learning, *Pulliam* and *Dunford* (1980) used the analogy of an investor and stockbroker. An investor could give his broker strict instructions to buy stock A, hold it for six months, then sell it and buy stock B, hold it for two years, and so on.

This strategy might work if the stock market were perfectly predictable and the investor could see the future. However, if the stock market were unpredictable and other, more favourable opportunities for investment presented themselves, the broker would be helpless without further instructions from the investor. This situation is analogous to the set of genes the has constructed an organism to respond to environmental contingencies in certain inflexible ways. If the environment remains the same as the one in which the organism evolved (was selected for), the programmed responses to contingencies should suffice to guarantee reproduction of the genetic material. But, if the environment changes, then the original program for survival (modes of adapting to the environment) might not be successful and, if so, the genes will be selected against and perhaps disappear from the population. In the example of wasp behaviour given above, it could be said that the wasp's genetic programming was not prepared for the environmental changes brought about by the experimenters. An investor could use a different strategy. He could give his money to a broker and instruct him to use his own experience and expertise with the stock market and invest the money as he saw fit, seeking the maximum return. If the broker were astute and knowledgeable, the investor would stand a better chance of doing well, in face of an unpredictable market, using this strategy rather than the inflexible instruction described previously. This situation is analogous to a set of genes that has constructed an organism with a nervous system that has the ability of learn and very its made of adaptation to match changing environmental contingencies.

If the nervous system learns well and applies its knowledge astutely, the chances of the genetic material being passed on to future generations is enhanced. Throughout this text we will see how learning builds on a base of innate reflexes, and allows each animal to face the challenge of survival with the inherited predispositions handed down through the eons from generations of ancestors, plus the unique knowledge that it may have acquired in its own lifetime. We will see how animals learn to ignore stimuli that seem unimportant, how they learn to anticipate changes in their environment, to prepare for these changes and, where possible to modify their environment and profit from the experience of being able to do so. However, before going on to a consideration of the simplest type of learning–habituation–we will first define learning more formally and specify some of the terms used in the experimental study of learning.

## Learning Variables

The variables that is of interest in the experiment, the one that is manipulated by the experimenter, is termed the independent variable. The term independent is used because the experimenter decides whether or not to administer the variable and what levels of the variable to administer. In the two examples given previously, the different types of automobiles would be the independent variable in the collision-injury study, and the presence or absence of the drug would be the independent variable in the maze-learning study. The experimenter is interested in the effects of the independent variable on some other event. In order to determine these effects, the experimenter varies the independent variable while holding other variables constant. In order to assess the effect of the independent variable, some measure of the event of interest must be taken. The variable used as a measure is termed the dependent variable. It is termed dependent because changes in this measure dependent upon changes in the independent variable; it reflects the influence of the independent variable. In the examples above, some measure of damage to the mannequin might be the dependent variable in the automobile study, whereas the number of errors made in the maze might be the dependent variable in the drug study.

## The Generality of Learning

In the chapters the follow, we will be discussing particular areas of research in some detail. In each case, we will use experimental data to support the general conclusions that are reached. In most causes, cases, the experimental subject cited will be the rat because of the widespread use of this animal as a subject. But is it reasonable to base general conclusions on the behaviour of a specific organism? Would these same generalities apply to a sea slug, an insect a chimpanzee, a fish, or a college sophomore? In many cases, the differences in behaviour that would be observed by using another subject would be trivial.

The purpose of this text is to develop an understanding of the basic principles of learning, and it is the contention of the author that most of these principles are applicable across species. Although our major purpose is to develop general conclusions that apply to virtually all organism, we will also examine a variety of situations in which the nature of the organism clearly interacts with the nature of the task. Some organisms, for example, easily acquire an active running response to avoid an aversive situation, but have difficulty learning to be passive and sit still; other organisms show the opposite patterns. Similarly, it will be seen that many birds tend to use the colour of food as a cue in learning tasks in which rats respond to taste as a cue.

Even considering only one kind of organism–pigeons, for example–we will see that the precise way in which a pigeon pecks at a signal for a reward differs depending upon whether the reward is water or grain. Thus, although we will find many examples of generality in learning, we must also remember that each species evolved by adaptation to a particular niche in the environment and that both the structure and the brain of the organism must be somewhat specialized for the ecological niche to which it is adapted. For example, carnivores and herbivores have evolved quite different feeding behaviours. It would not be surprising to find that a cat (a carnivores and herbivores have evolved quite different feeding behaviours.

It would not be surprising to find that a cat (a carnivore) and guinea pig (a herbivore) behave somewhat differently in learning tasks that use a food reward. However, although differing in detail, the general principles of learning remain similar in such divergent species. Similarly, casts tend to be solitary hunters or hunt in very small groups, whereas canids (dogs, wolves) tend to hunt in large groups. Again, it would be surprising if the long evolutionary history that adapted these animals to such different social behaviours did not also lead to differences in brain structure that would influence some types of laboratory learning tasks. When an animal is taken out of its niche and brought into the laboratory, the experimenter must be careful to temper any conclusions about learning ability or learning principles with a considerable of how the learning task might relate to the animal's behaviour in its natural environment. Later in the text we will consider in some detail how the evolutionary histories of various species may interact with the degree and type of learning that they exhibit in the laboratory.

According to Wallace (1973) learning is the adaptive change in behaviour that result from the individual experience and probably associated with physical changes in the central nervous system. It may function with innate pattern but may ultimately be separable from these patterns. Many undergraduates who enroll in learning courses are interested in human earning and they see little value in the study of animal learning. There are, however, several reasons to pursue such investigations. One such reason is that there may be similarities in principles and/ or physiological processes underlying learning in humans and other animals. We will see many instances of this in subsequent chapters. For example, we will see how research on Pavlovian conditioning in animals has provided a way of understanding the learning of emotional responses in humans. It is quite likely that there are substantial similarities if not identities in the physiology, neurochemistry, and learning principles governing emotional behaviour across the animal spectrum.

Knowledge of these similarities helps in the understanding and treatment of phobic behaviour in humans, and it should soon help in the understanding of how emotional stress contributes to disease states such as hypertension and cancer. We shall also see how research in Pavlovian conditioning and food aversion learning in animals may have implications for the understanding of difficulties that cancer patients face when they undergo chemotherapy. Another example that may be cited in this regard is the widespread use and effectiveness of behaviour modification techniques in treating various behaviour problems.

The basic understanding of these techniques was derived from animal research. Many other examples of direct applications of animal learning to human behaviour will become apparent throughout the text. One such example is the relevance of the study of punishment and aversive control in animals for the understanding of the relative effectiveness of reward and punishment in shaping the behaviour of children. A second example is the relevance of the study of incentive relativity for the understanding of disappointment in humans. Even the use of language–very likely unique to humans–as an instrument to produce a particular outcome may follow some of the same principles as instrumental learning in animals, where they learn that certain behaviours will produce particular changes in the environment.

In other words, although the acquisition and many of the uses of language for congnitive processes may be unique to humans, the use of language as a tool in the service of notives and emotions may be similar to instrumental behaviour in lower animals. If these examples are relevant for the behaviour of humans, why not study them directly in humans? In some cases that is possible after the relevance of the animal research is perceived. In other cases it is not possible because of the importance of the principle of control. We have seen how extraneous variables must be controlled if the effectiveness of a particular variable on behaviour is to be understood.

In many instances it is not possible to control factors that might influence learning in humans–not possible because of practical reasons and/or ethical reasons. For example, if the effectiveness of a particular treatment on learning is to be investigated, how is it possible to ensure that two groups of humans come into the experiment with the same past experience: It is not possible. Also, human know when they are in an experiment and this knowledge itself may influence the outcome of an experiment. (There are somewhat related problems in animal research, but they are much easier to control). Thus, research that may have direct relevance to human behaviour is often best done with animals. A second reason for investigating learning in animals is that it may help us to understand what is unique about human learning. Although there are many examples of similarities between human and animal learning, it is clear that there must also be many differences. Other animals, for example, do not have differences.

Other animals, for example, do not have the language abilities that humans' possesses–even the apparent rudimentary language structure seemingly demonstrated in chimpanzees is under question. There are probably many differences in cognitive abilities between lower animals and humans, and the extent of these differences will become clearer the more we know about animal learning. That is, the knowledge we have of animal learning may serve as background against which the unique aspects of human learning of human learning may be perceived more clearly.

## Learning through the Conditioned Response

Animals with well-developed nervous systems can learn to behave in particular ways. Learned behaviour differs from innate behaviour in that the animal develops new responses to situations, and retains these responses for an extended period of time. Many experiments have been preformed to see if protozoans can learn. None of these experiments, however, has been able to show definitely that protozoans can learn. A few of these experiments are open to various interpretations, however, and some biologists might argue that some form of learning took place. We would be surprised if there were sharp line between animals that can learn and those that cannot learn. We might predict instead that there would be a spectrum of abilities from totally innate behaviour to the highest forms of reasoning. Sometimes, even in the vertebrates, it is difficult to know if learning actually has occurred. For instance, does a salamander learn to swim, or does it swim spontaneously when it reaches a certain age and stage of development?

In an experiment to answer this question, a group of salamanders were anesthetized at the age just before swimming movements normally begin. A control group was allowed to develop until they swam normally. When the experimental salamanders were allowed to recover from the anesthesia, they immediately swarm normally. These experiments therefore showed that swimming movements in the salamander are the result of normal development. Apparently learning has nothing to do with their swimming. One of the best-understood types of learning is the conditioned response, also called a conditioned reflex.

A conditions response involves the substitution of one stimulus for another. That is, assume stimulus A produces response B. If stimulus C can be made to produce response B, then one stimulus has substituted for another and a conditioned response has been formed. Experimenters have tried to develop conditioned responses in many types of animals. One of the most interesting experiments was performed on planarians. In the early 1950's, Robert Thompson and James McConnell, working at the University of Texas, developed a conditioned response in planarians. First, they shined a strong light on a planarian. Several seconds later, they administered a mild electric shock.

The planarian's normal response to the light was to stretch. Its normal response to the shock was to contract or turn its head. This sequence, in which the light was followed by the shock, was

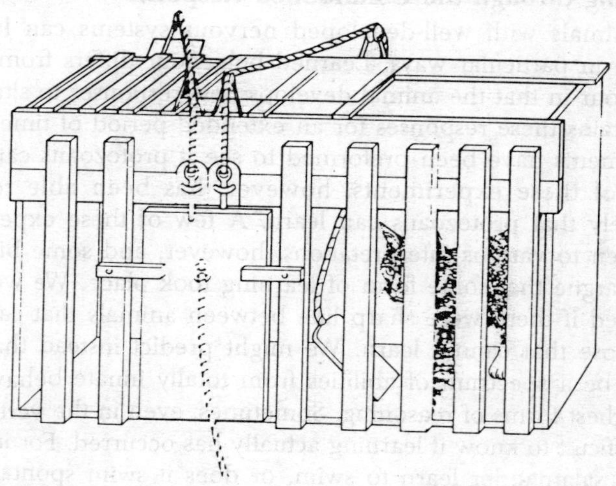

*Fig. 4.1. A cat in one of Thorndike's puzzleboxes.*

repeated about 100 times. Soon the response to the light alone was about the same as though the electric shock had also followed. This sequence is outlined. A "trained" planarian would show a shock response when the light was turned on 23 out of 25 times. It soon forgot its lesson if not retained periodically. In later experiments, trained planarians were cut in half. Each half was allowed to grow. The tail half grew a head; the head half grew a tail. The regenerated planarians were then conditioned to the light-shock response. It was found that both types of planarians, those regenerated from the head-half and these regenerated from the tail-half, learned quicker than planarians being conditioned for the first time.

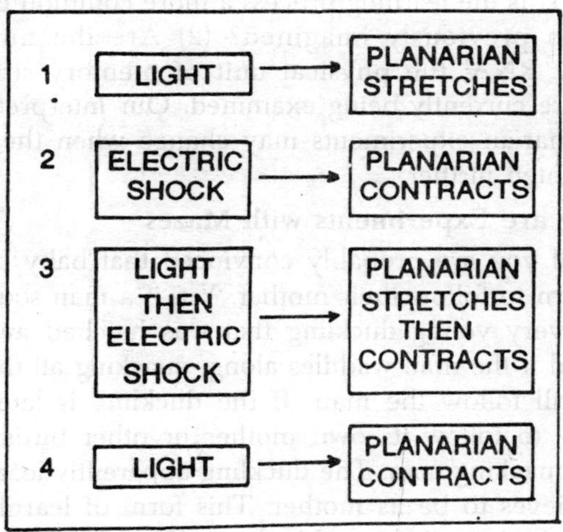


*Fig. 4.2. The conditioned response in planarian.*

Apparently some memory storage had occurred in planarians regenerated from tail pieces was well as in those regenerated from head pieces. You may want to refer back to review some of the regeneration experiments performed on planaria. These results suggested additional experiments to other biologists. Also taken into account were previous experiments on rabbits that had shown that RNA was probably associated with memory. Planarians were trained, cut in half, and allowed to regenerate as before. This time, however, the regeneration was performed in a solution ribonuclease (rye-both-New Kleeays), an enzyme which destroys RNA. When the regenerated planarians were retained, those regenerated from the tail-half learned about the same as a planarian which had never been trained, but those regenerated from the head-half learned at a significantly faster rate. It is not yet clear why there is this distinct difference in reaction between the two types. A second experiment with very significant results was designed to test the effect of a special type of cannibalism on memory storage. Trained planarians were chopped into small pieces. These pieces were then fed to a group of untrained planarians.

The control group consisted of planarians which were fed pieces of untrained planarians. The interesting result: the planarians that were fed bits of trained planarians learned better than the control group. These experiments on planarians have raised several questions: (1) Is the learning process a more common characteristic of life than previously imagined? (2) Are the nucleic acids, particularly RNA, the physical unit of memory storage? Such questions are currently being examined. Our interpretation of the trained planarian experiments may change when these questions are investigated further.

**Imprinting are Experiments with Mazes**

Most of you are probably convinced that baby birds do not have to learn to follow their mother. Yet, if a man squats down in front of a very young duckling that was hatched away from its mother, and if the man waddles along, quacking all the while, the duckling will follow the man. If the duckling is later given the opportunity to follow its own mother or other birds, it will still follow the quacking man. The duckling apparently learns to follow what it believes to be its mother. This form of learning is called imprinting. In investigating this form of learning further, it was found that a duckling will learn to follow a large coloured box with a ticking clock inside if the box is the first thing it observes. When the coloured box was pulled along on a wire, the duckling followed along, preferring the box to its own mother.

Imprinting is an unusual form of learning. These experiments certainly show that early attraction of a duckling to the mother bird is not instinctive–it is learned by imprinting because the mother bird is usually the first object that the duckling sees. For many years, a favourite way to investigate the learning ability of animals has been with the use of a maze. A maze is a series of passages in which an animal must choose between alternate paths. If an animal makes the right series of choices, he will be rewarded with food. If he makes the wrong series of turns, he may be

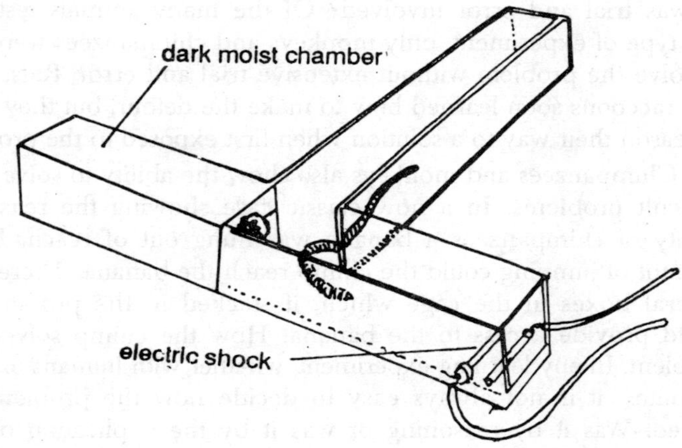


*Fig. 4.3. An earthworm in a T-maze*

punished, with a mild electric shock, for example. By being put through the maze again and again, an animal may learn to make the turns that result in reward rather than punishment. The simplest kind of maze is the T-maze where only one choice is involved.

Flatworms and earthworms can learn to make the "correct" choice of turns in this maze. Earthworms, for instance, are given the choice either of entering a dark, moist chamber or of receiving an electric shock. The earthworms took about 200 trials in the maze to learn to make the correct turn. After the earthworm had learned, it could make the correct turn in 90 per cent of the cases. If it had not learned, it would have been just likely to turn right as to turn left. A number of kinds of insects learn mazes quite rapidly, but even among insects learning ability differs. Ants are better learners than cockroaches, for example.

## Reasoning is Characteristic of Man and Some Primates

The highest form of learned behaviour, and the one in which man excels, is reasoning. Reasoning can be used to solve complex problems without resorting to trial and error, the method used in maze learning. Reasoning ability can be studied with the "detour problem." In such a problem, and animal must follow an indirect path to get to a desired object, usually food. An animal must move away from the food before he can reach it. In experiments with the detour problem it is important to consider the following question: Did the animal solve the detour during its initial attempts or was trial and error involved? Of the many animals tested in this type of experiment, only monkeys and chimpanzees were able to solve the problem without extensive trial and error. Rats, dogs, and raccoons soon learned how to make the detour, but they failed to reason their way to a solution when first exposed to the problem.

Chimpanzees and monkeys also show the ability to solve more difficult problems. In a now classic case showing the reasoning ability of chimpanzees a banana was hung out of reach. By no amount of jumping could the chimp reach the banana. There were several boxes in the cage which, if stacked in the proper way, could provide access to the banana. How the chimp solved this problem. In any learning experiment, whether with humans or other primates, it is not always easy to decide how the problem was solved. Was it by reasoning or was it by the application of past experience? Or is reasoning actually an accumulate effect of experiences? The answers to these questions are at the heart of exciting work in the field of behaviour. How man and other animals learn is an important phase of scientific research.

Man is capable of much higher levels of reasoning than the other primates. He can learn to use symbols such as letters and numbers at a comparatively early age. These means of communication have made man's knowledge and civilization possible. That man's responses to letter and number symbols are forms of learned behaviour is made quite clear from the fact that no baby is born with the ability to read. He must be taught reading, writing, and arithmetic. These are basic skills for which only the ability to learn is innate. When he has learned these skills, he can then make further use of his extraordinary mental capabilities. The ability to reason, one of man's most cherished possessions, is also one of his least understood possessions.

## Forms of Learning

Animal behaviourists have used many different procedures in studying learning. We shall next consider some of these. A description of the various forms of learning is analogous to the observation and description that provides the starting point for the study of all animal behaviour (*Tinbergen,* 1963). Many of the distinctions to be made among different "forms" of learning are based on differences in the procedures used in the study of learning. The fact that the procedures used differ need not necessarily imply that the processes underlying these different forms of learning are either similar or different. There may be one, two, or many kinds of processes underlying different forms of learning.

## Habituation

*Habituation* may be defined as the decrease in probability or amplitude of a response, occurring when a stimulus which elicits that response is presented repeatedly. In an excellent discussion of the phenomenon of habituation, *Thompson* and *Spencer* (1966) listed nine characteristics of habituation that are manifested in most systems. We shall consider a few of these. If, after repeated stimulus presentation and habituation of the response, the stimulus is no longer presented, the response tends once again to be elicited by the stimulus. Within limits, the longer one withholds the stimulus the greater is the likelihood that the response originally elicited by the stimulus will again be elicited by it. This phenomenon is called *spontaneous recovery*. It may require a few minutes to several days, depending on the conditions. If a repeated series of habituation training trails and spontaneous recovery trails is given, the habituation occurs progressively more rapidly (***potentiation of habituation***).

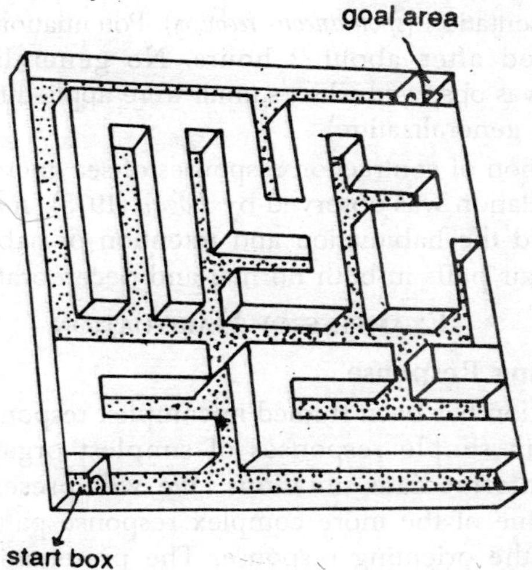


*Fig. 4.4. A rat may learn a maze by trial-and-error processes, initially making many wrote turns and entering numerous clusde-sac. Gradually the rat makes fewer error, until, after a number of trials, it completes the maze without any error. Also, a rat may perform with fewer error when it has been given prior exposure to the maze with nor rewards present, a process called latent learning.*

Habituation of a response to a given stimulus shows generalization to other similar stimuli (*stimulus generalization*). Thus, if an animal habituates a response to a tone of a given frequency, it is likely that responses to tones of frequencies close to the training stimulus will also show habituation. The more dissimilar the test tone from the habituation tone, the less would stimulus generalization be apparent. Finally, presentation of another (usually very intense) stimulus results in a sudden recovery of the habituated response (*dishabituation*). In order to be considered true habituation, the response decrement must clearly be the result of changes in the central nervous system rather than of *adaptation* in the sensory system or fatigue in the effectors. When a puff of air is applied to the gills of a horseshoe crab, it responds with a movement of its telson (tail). *Lahue, Kokkinidis,* and *Corning* (1975) studied the characteristics of habituation of the telson reflex, monitoring the muscle movements electrophysiologically. They found a clear habituation of the telson reflex to repeated presentations of the air-puff stimulus. A brief squirt of saline from a syringe aimed at the site of habituation resulted in dishabituation. Initial levels of responsivity were normal after an interval of 1 hour without stimulus presentation (*spontaneous recovery*). Potentiation of habituation was observed after about 2 hours. No generalization of the habituation was observed when stimuli were applied to other regions (no stimulus generalization).

Habituation of contraction responses of sea anemones to water-stream stimulation was observed by *Logan* (1975). *Ratner* and *Gilpin* (1974) studied the habituation and retention of habituation of the response to air puffs in both normal and decerebrate earthworms.

## Examples of Habituation

### The Orienting Response

Habituation has been studied in complex responses of complex organisms, in simple responses of simplest organisms, and in numerous intermediate situations. We will present only a few examples. One of the more complex response patterns occurs in the case of the orienting response. The presentation of a novel stimulus of moderate intensity will elicit a variety of behavioural and physiological changes. For example, the organism will tend to orient toward the stimulus, there may be a decrease in its heart rate, a decrease in skin resistance, a decrease in respiration, changes in the construction of peripheral blood vessels, a change in muscle tone, and changes in the electrical activity of the cerebral cortex (Sokolov, 1963). If the stimulus is repeated a few times with no untoward consequences to the organism, then all of these responses will habituate, although perhaps at different rates. The habituation of one component of the orienting response–the change in skin conductance.

The presentation of a loud clicking sound on the first two trials elicits a large degree of skin conductance caused by activity of the sweat glands, which, in turn, reflects the activity of the sympathetic nervous system. As the loud click is repeated, the degree a change in skin conductance diminishes until eventually there is little response to the noise. The subjects in this experiment

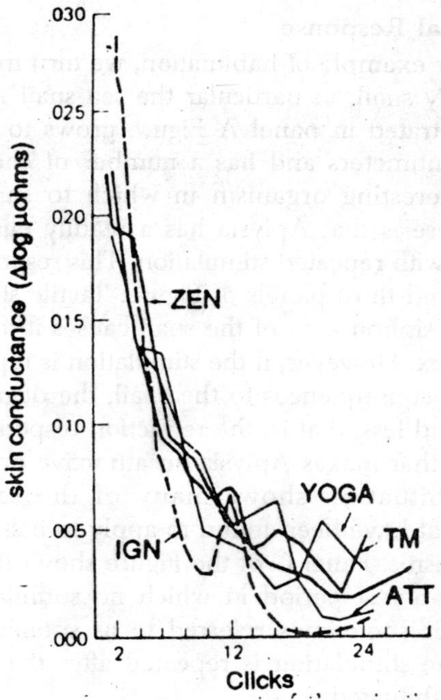


*Fig. 4.5. Habituation of one component of the orienting response, skin conductance, as a function of the number of clicks presented. Different groups of subjects practiced their usual meditational techniques, Zen, Yoga, or Transcendental Meditation (TM), or were instructed to ignore (IGN) or attend (ATT) to the loud clicks. The first few loud clicks produced a substantial increase from baseling (.000) in skin conductance (a measure of sympathetic nervous system activity) in all groups. Subsequent clicks produced less and less of an effect. The rate of habituation was not affected by the different states of consciousness.*

were humans separated into five different groups. Three of the groups were composed of highly experienced meditators using different techniques. These groups were instructed to engage in their standard meditating practice during presentation of the clicks. One of the other two groups was instructed to ignore the clicks (IGN) and the fifth group was instructed to attend to the clicks (ATT). Rate of habituation was approximately equal in all groups; apparently an inconsequential stimulus is an inconsequential stimulus in any of a variety of states of consciousness. Interestingly, Leaton and Jordan (1978) showed that habituation in rats to a loud noise, accomplished while they were awake, transferred to the sleep state. That is, rats that had been previously habituated were less likely to be aroused from sleep by the loud noise.

**Gill Withdrawal Response**

For another example of habituation, we turn from humans and rats to the lowly snail, in particular the sea snail *Aplysia california*. This snail, illustrated in panel A Figure grows to an adult length of about 20 centimeters and has a number of characteristics that make it an interesting organism in which to study habituation. The first of these is that Aplysia has a readily elicitable response that habituates with repeated stimulation. This response is illustrated in the second and third panels of Figure. Tactile stimulation of the mantle shelf of siphon area of the snail causes it to retract its gill– a defensive reflex. However, if the stimulation is repeated and there are no adverse consequences to the snail, the degree of retraction becomes less and less, that is, the retraction response habituates. A second feature that makes Aplysia an attractive organism to study is that its habituation shows many of the same functional relationships that have been found to apply to habituation in more complex organisms. Panel C of the Figure shows that the response reappears after a rest period in which no stimulation is applied. This recovery is sometimes referred to as spontaneous recovery. Note that as the stimulation is repeated after the rest period, the response is rehabituated.

At first glance, this effect of a rest interval leading to the recovery of the response might indicate that the habituation is really due to a fatigue process and process and is not the result of learning. However, the next panel in Figure illustrates the results of one of several procedures that may be used to show that habituation is not due to fatigue. This fourth panel (D) shows that

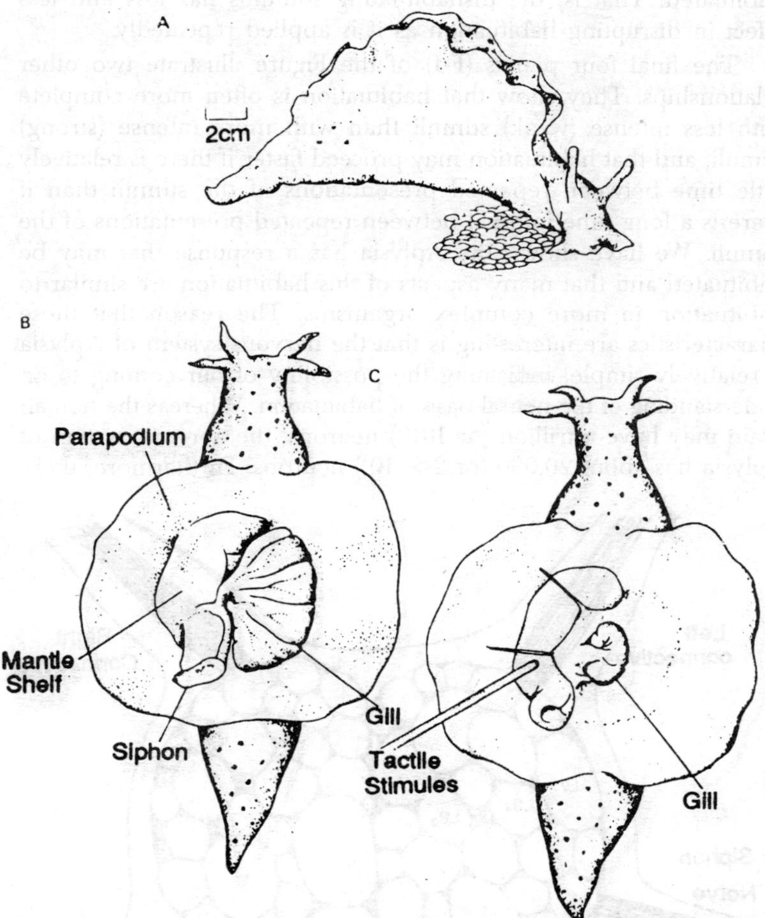


*Fig. 4.6. The sea snail Aplysia californica. Tactile stimulation of the mantle shelf or siphon causes reflex retraction of the gill. This response habituates with repeated stimulation.*

habituation can be counteracted if the animal experiences another strong stimulus. For example, if Aplysia is touched somewhat roughly in the head and then the mantle shelf is stimulated, the previously habituated gill retraction response will reappear. This return of the response is referred to as dishabituation, and it indicates that the animal's muscle system is still capable of making the habituated response–it has not been lost because of fatigue. The next panel of Figure shows that dishabituation itself may be habituated. That is, the dishabituating stimulus has less and less effect in disrupting habituation as it is applied repeatedly.

The final four panels (F-I) of the Figure illustrate two other relationships. They show that habituation is often more complete with less intense (weak) stimuli than with more intense (strong) stimuli, and that habituation may proceed faster if there is relatively little time between repeated presentations of the stimuli than if there is a long time interval between repeated presentations of the stimuli. We have shown the Aplysia has a response that may be habituated and that many aspects of this habituation are similar to habituation in more complex organisms. The reason that these characteristics are interesting is that the nervous system of Aplysia is relatively simple, indicating the possibility of our coming to an understanding of the neural basis of habituation. Whereas the human brain may have a trillion (or 1012) neurons, the nervous system of Aplysia has about 20,000 (or $2 \times 10^4$) neurons. Furthermore, there

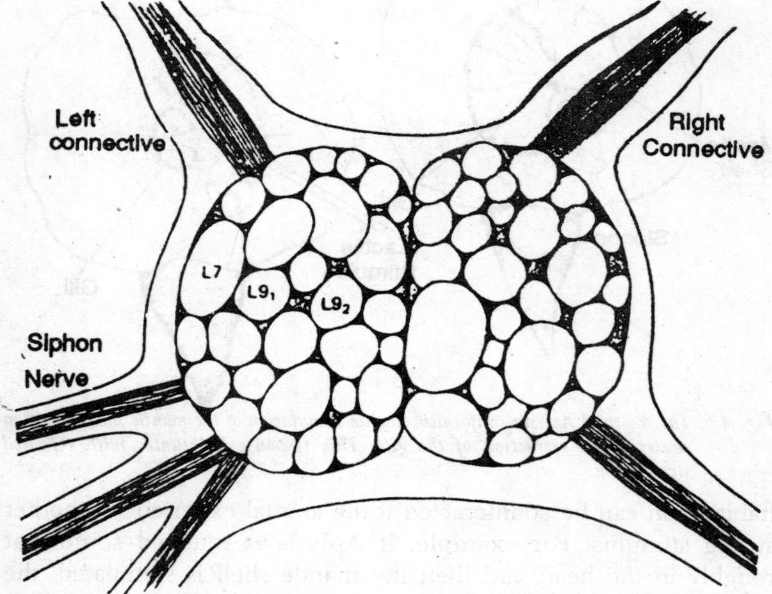


*Fig. 4.7. Schematic drawing of nerve cells in the abdominal ganglion of Aplysia. Many of the neurons are large and are identifiable in regard to location and function in different animals. The neurons labeled L7 and L9 (left 7, etc.) are involved in the gill withdrawal reflex. Also shown are nerve fibers connecting the abdominal ganglion to other parts of the nervous system.*

neurons are organized into nine separate groups (ganglia) each containing about 2,000 neurons. Many of these neurons are large and recognizable as particular cells occupying distinct locations and having distinct functions that are invariant in each individual. Some of this structure may be seen in the map of the abdominal ganglion of Aplysia presented in located in this abdominal ganglion (L7 is one of these motor neurons).

## The Neural Basis of Habituation in Aplysia

What happens at the neural level when a response becomes habituated? A long series of studies by Kandel and his colleagues has provided a substantial answer to this question in the case of the gill withdrawal response in Aplysia. A simplified version of their discoveries is presented in Figure. When the mantle shelf of siphon are touched, an impulse is initiated in sensory neurons. These sensory neurons influence. The first few times the sensory neuron is activated, there is substantial activation of the motor neurons and withdrawal of the gill. As the stimulus is repeated, there is less and less activation of the motor neuron. Does this mean that the motor neuron is becoming fatigued? No–because

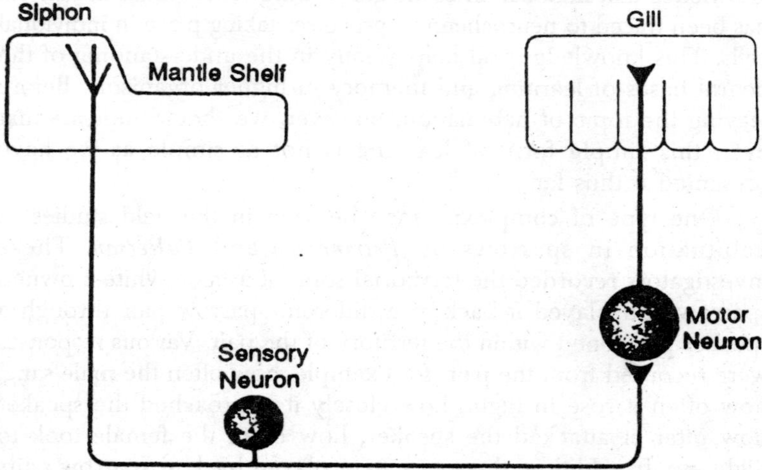


*Fig. 4.8. Simplified "wiring diagram" of the gill withdrawal response in Aplysia. The sensory neuron detects tactile stimulation of the siphon or mantle shelf area. The sensory neuron then releases a chemical transmitter (acetylcholine) from the axon terminal. The transmitter crosses the synapse and induces activity in the motor neuron (L7) which acts to retract the gill. With repeated stimulation, the amount of transmitter released becomes less and less.*

direct stimulation of the motor neuron shows that it is still capable of producing the complete gill withdrawal response. Is there less activity in the sensory neuron with repeated stimulation? No-Kandel's work has shown that the response in the sensory neuron is essentially undiminished by repeated stimulation. It seems that the change that underlies habituation occurs in the region of the synapse between the two cells. Kandel has been able to show that this change involves the release of less and less transmitter substance by the sensory nerve as stimulation is repeated. With less transmitter substance released, there is less stimulation of the motor nerve and less of a gill withdrawal response. Is less transmitter substance being released because the supply is exhausted by repeated stimulation?

No–the phenomenon of dishabituation indicates that this is not likely to be the case and, in addition, Kandel has shown that the release of less transmitter substance is an active process that results from other changes in the cellular chemistry of the sensory neuron. Unfortunately, we will not be able to delve further into this cellular chemistry in this text. We will have to be satisfied with the knowledge that habituation of the gill withdrawal response in Aplysia has been traced to neurochemical processes taking place in individual cells. This knowledge will help greatly in the understanding of the neural bases of learning and memory in higher organisms. Before leaving the topic of habituation, however, we should indicate that even this simple form of learning is not as simple as we have presented it thus far.

One type of complexity may be seen in the field studies of habituation in sparrows by *Petrinovich* and *Patterson.* These investigators recorded the territorial song of a male white-crowned sparrow and played it back to a different sparrow pair through a speaker positioned within the territory of the pair. Various responses were recorded from the pair–for example, how often the male sang, how often it rose in flight, how closely it approached the speaker how often it attacked the speaker, how often the female took to flight etc. In addition, the songs were played back to sparrow pairs in different brooding conditions, that is, some were brooding eggs, some hatchlings, and some fledglings. The results of these studies indicated that habituation generally occurred, but the degree of habituation depended upon the brooding condition, the nature of the response, and the manner in which the song was presented.

The point is that all responses may not habituate at equal rates and that the rate at which a given response habituates may depend on motivational states (here represented by brooding conditions) of the animal.

**Different Processes**

A second complexity is that all response changes that superficially appear to reflect the same process of habituation may not, in fact, reflect the operation of the same mechanism. For example, if rats are exposed to a loud click it will elicit a startle response that may be measured in terms of how much they jump. As the click is repeated, it produces less and less of a startle response. In other words, the startle response habituates. If a rat is exposed to a maze it has never experienced before, it will engage in a great deal of activity "patrolling" the maze. As the animal is given repeated exposure to the maze, its tendency to enter the various branches declines, that is, it seems to habituate.

Although behaviourally there would seem to be the same process occurring in these two situations–decline of a response with repeated exposure to the stimulus–there is evidence that two different brain mechanisms may be responsible for the two declines. Leaton has found that damage to the hippocampus, a part of the brain implicated in memory processes in humans, prevents the decline in exploration seen with repeated exposure to the maze, but such damage has no effect on habituation of the startle response. That is, the startle response habituates in about the same way in brain-damaged rats and in normal rats, but the rate of decline in maze exploration is quite different in the two groups. Thus, there may be a number of different brain mechanisms controlling different kinds of habituation.

In fact, it has been argued that maze explora-relations and the location of objects in space–and when the map is completed, exploration ceases *O'Keefe* and *Nadel* have also hypothesized that the hippocampus in necessary for the formation of these cognitive maps, and that without an intact hippocampus the animal cannot learn (or remember) spatial relations. Thus, exploration continues unabated. From this point of view, then, the apparently similar habituation curves of the startle response and maze activity would really represent two quite different processes–the learning of spatial relations in the case of the maze, and perhaps the learning that the loud click is unimportant in the as of startle.

### Short and Long-Term Habituation

As a final complexity, we shall mention the possibility that there are two basic types of habituation: short-term and long-term. One way of demonstrating the difference in them is to conduct a relatively brief series of trails and them is to conduct a relatively brief series of trails and then check for recovery from habituation. Recovery usually occurs fairly quickly and completely. When, however, a longer series of trails is given, there is less recovery, and habituation is maintained over a longer time period. Another way of demonstrating this is to present the stimulus to be habituated at two different interstimulus intervals (ISI) to two different groups of animals. For example, one group could be presented a stimulus every two seconds, whereas a second group could be presented a stimulus every 16 seconds. When this is done, the short ISI group.

When however, these same two groups are tested some time after the initial habituation session, then the long ISI group might be likely to show less recovery from habituation than the short ISI group. It is possible that these two types of habituation may be related to difference in short-term and long-term memory in humans. Given that there are two types of habituation, it is necessary to postulate that there are two different neural processes responsible for them?

Some research by Kandel (1979a) indicates that the answer to this question is "yes and no." That is, in both short-term and long-term habituation, the sensory neuron of Aplysia becomes less effective in stimulating the motor neuron and the basis of this effect is that there is less transmitter substance released at the synapse. Thus, the same basic mechanism seems to be involved in both types of habituation. However, there must also be a difference since synaptic activity is depressed for longer time periods in long-term than in short-term habituation. Kandel speculates that there may be some structural changes in the presynaptic terminal of the sensory neuron that result from the extended habituation training. These structural changes then contribute to the longer-lasting effects of habituation. However, the effects of these structural changes are reversible, as we shall see when we discuss our next topic sensitization.

## Sensitization

### Incremental Sensitization and Dishabituation

Sensitization refers to the augmentation of a pre-existing response by a strong, usually noxious, stimulus. There are two somewhat different ways in which the term sensitization is used. The first use is in referring to the enhancement of a response elicited by a stimulus with repeated presentations of that stimulus. We shall refer to this as incremental sensitization. It is clear from the definition the incremental sensitization describes a behavioural effect that is the opposite of habituation. A period of sensitization of ten precedes habituation. That is, the behavioural effects of the second, third or forth presentation of a stimulus way be greater than the effects produced by the first presentation. Subsequent presentations may, however, lead to less and less of a behavioural effect. Habituation, therefore, may follow a period of incremental sensitization. The first panel shows an initial marked increase in the frequency with which a male three-spined stickleback fish (*Gasterosteus aculeatus*) bites another male introduced into its territory. After the fourth minute of intrusion, habituation of the attacks sets in. The second panel shows a similar process in the number of song sung by a male white-crowned sparrow when the song of another male is played in its territory. In this case, the frequency of singing did not fall below the initial level during the course of the eight trails in which the "intruder's" song was played.

The third panel shows changes in skin conductance in humans as a function of the intensity of sounds presented to them. Remember from our earlier discussion that skin conductance changes are one component of the orienting response. Figure shows clearly that the degree of sensitization depends upon the intensity of the stimulus presented. When loud sounds are presented, there is a period of incremental sensitization that precedes habituation. A second way in which sensitization is often seen in behaviour occurs when a stimulus, usually intense or noxious, is presented just prior to the presentation of an already habituated stimulus. In this use, the term sensitization is synonymous with dishabituation. It is as if the intense stimulus primes the animal (sensitizes it) to respond to any change in the environment. By thus priming the animal, the new stimulus undoes the effects of the previous habituation period.

It is easy to imagine the potential adaptive value of the sensitization-as-dishabituation process. Once an animal has habituated to a stimulus, it is in a vulnerable state should that stimulus suddenly become dangerous. The occurrence of dishabituation following the experience of another stimulus that is intense or noxious may serve to reduce this potential vulnerability. That is, the conditions under which habituation occurred may no longer be the same, and it may be safest for the animal if its nervous system is constructed so as to return to initial conditions (dishabituation) when the environment is changed.

## Dual Process Theory

What accounts for functions where responsivity tends at first to increase before declining? *Richared* Thompson and his colleagues have postulated that there exist two independent process that are

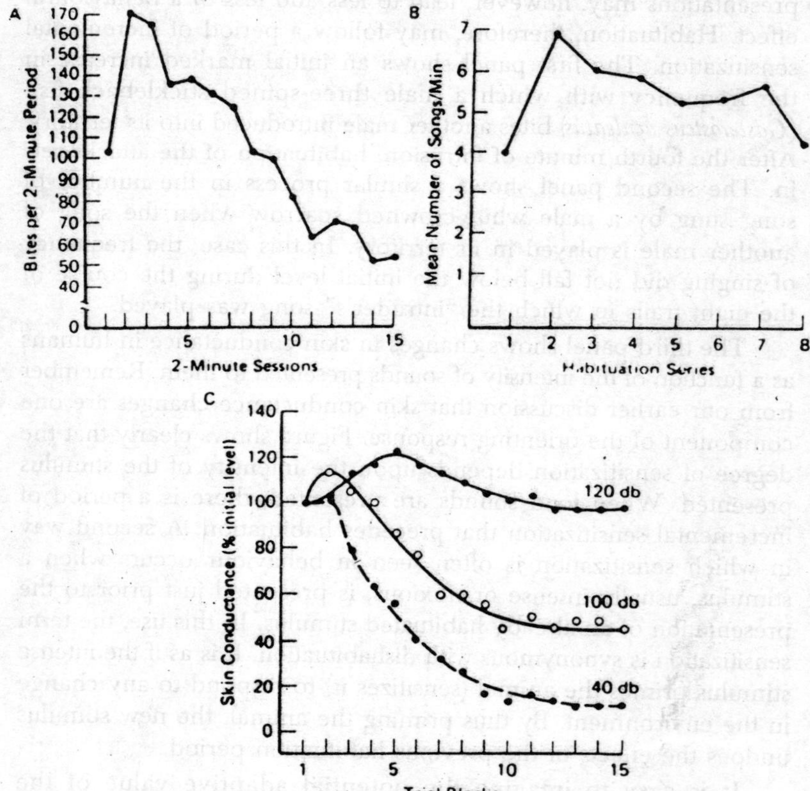


*Fig. 4.9. Examples of sensitization. Panal A shows an initial increase in the bitting attacks of a three-spine stickleback fish toward a male conspecific introduced into its territory. Panal B shows initial sensitization in the number of songs sung by a male white-crowned sparrow when the song of another male is played in its territory. Panal C shows changes in human skin conductance as a function of the intensity of auditory stimuli.*

aroused by the presentation of a stimulus: habituation and sensitization. That is, each presentation of a stimulus has a tendency to sensitize an animal to future presentations of that stimulus so that such presentations will produce an enhanced response. At the same time, each presentation of a stimulus leads to an increment in habituation such that future presentations of the stimulus will tend to elicit a smaller response. The actual behaviour that occurs represents the outcome of these opposed processes. The two panels show hind limb flexor responses elicited by shock in cats. The cats in these experiments were prepared so that shock influenced only sensory neurons in the spinal cord, and the response of the cat's leg was controlled only by motor neurons in the spinal cord. The brain was not involved.

In addition to the responses obtained in this experiment (represented by the solid lines), the Figures also shows the hypothesized course of the two processes of habituation and sensitization (represented by the dashed lines). The behavioural data is assumed to reflect the interaction of these two processes. The first panel of the Figure shows data and hypothetical processes

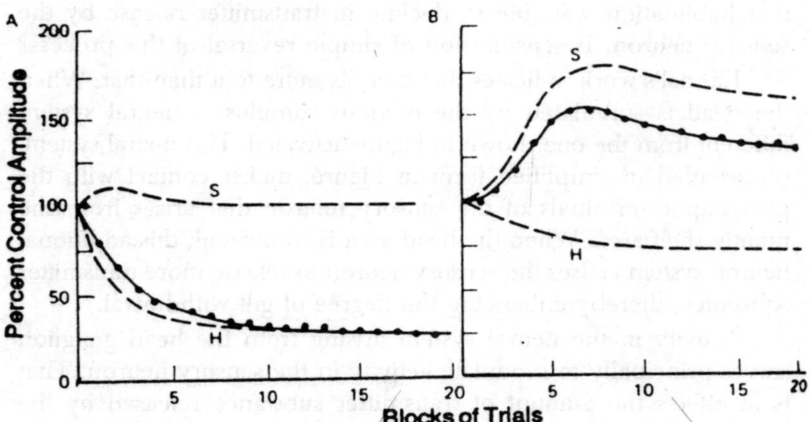


*Fig. 4.10. Habituation and sensitization of hind limb flexor responses in cats. The data in panal A were obtained when a low intensity showk was used; the data in panel B were obtained with a high intensity shock. The solid lines in the two panels represent changes in the degree of flexor response. It is apparent that habituation predominated with low intensity shock, and sensitization with high intensity shock. The dashed lines in each represent theoretical processes of sensitization (S) and habituation (H) that underlay the obtained results. Sensitization is assumed to act always to offset habituation. The more intense the stimuli, the more pronounced the sensitization component.*

when the intensity of the shock was low. In this case there was no evidence of incremental sensitization. The assumption is that at low intensities there is little sensitization and the habituation process predominates. The second panel shows the results and hypothetical processes when a more intense shock was used.

In this case there was substantial incremental sensitization and little evidence of habituation. The inference drawn from these and similar data is that there is always some balance between habituation, which predominates when stimulus intensity is low, and sensitization which predominates when stimulus intensity is high. Some evidence in support of this dual process interpretation of habituation and sensitization has been obtained at the physiological level. For example, Kandel and his colleagues have studied sensitization in Aplysia. The found that a noxious stimulus applied to the head of Aplysia will greatly enhance a previously habituated gill withdrawal reflex. They have also found that this enhancement is brought about because of an increase in the amount of transmitter substance released by the sensory neuron. Recall that in our earlier discussion of habituation in Aplysia we stated that habituation was due to decline in transmitter release by the sensory neuron. Is sensitization of simple reversal of this process?

Kandel's work indicates that there is more to it than that. When the head is stimulated by the noxious stimulus, a neural system different from the one shown in Figure activated. This neural system, represented in simplified form in Figure, makes contact with the presynaptic terminals of the sensory neuron that arises from the mantle shelf area. When the head area is stimulated, this additional neuron system causes the sensory neuron to release more transmitter substance, thereby enhancing the degree of gill withdrawal.

Activity in the neural system arising from the head ganglion serves principally to modulate activity in the sensory neuron. That is, it affects the amount of transmitter substance released by the sensory neuron only when the mantle shelf is stimulated. Stimulation of the head without subsequent stimulation of the mantle shelf has relatively little effect on transmitter release and gill withdrawal activity. Thus, the evidence adduced by Kandel and his colleagues agrees with the model proposed by Thompson and his colleagues–sensitization and habituation represent two separate and distinct processes.

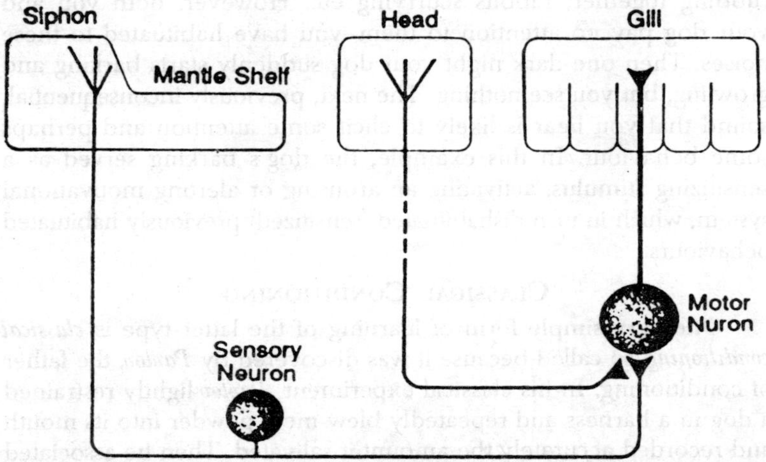


*Fig. 4.11. Simplified diagram of dishabituation circuit in Aplysia. Tactile stimulation of the head area produces activity in a neuron that terminates on the presynaptic axonal area of the sensory neuron. Release of a neurotransmitter (acetylcholine) from the sensory neuron when the mantle shelf area is stimulated. Thus, a second neural system mediates dishabituation.*

Other research supports this conclusion. For example, research by Kandel's group has indicated that the chemicals serving as transmitters are different in the two systems, at least in the case of Aplysia. Whereas acetylcholine is the transmitter released by the sensory neuron, serotonin is released by the neuron from the head ganglion. Also, Thompson and his colleagues have found evidence that there are two types of neurons in the cat spinal cord: one type that shows only habituation and another type that shows sensitization. Thus, the evidence seems sufficient to conclude that the decrements in behaviour that we see as habituation, and the increments in behaviour that we see as sensitization, represent two distinct processes served by different neural systems.

In general terms, we could think of habituation as representing a process of learning that a particular stimulus is not important–it has no consequences for the organism. However, if an aversive event occurs, then a motivation system is activated which enhances reactivity to environmental changes, even those that have been previously habituated. Perhaps one more example will help to clarify habituation and sensitization. Imagine that you make it a regular practice to walk your dog at night in the dark, wooded back yard. There are a lot of sounds in the might–tree branches rubbing together, rabbits scurrying etc. However, both you and your dog pay no attention to them–you have habituated to these noises. Then one dark night your dog suddenly starts barking and growling, but you see nothing. The next, previously inconsequential, sound that you hear is likely to elicit some attention and perhaps some behaviour. In this example, the dog's barking served as a sensitizing stimulus, activating an arousing or alerting motivational system, which in turn dishabituated (sensitized) previously habituated behaviours.

## Classical Conditioning

One very simple form of learning of the latter type is *classical conditioning,* so called because it was discovered by *Pavlov,* the father of conditioning. In his classical experiment, *Pavlov* lightly restrained a dog in a harness and repeatedly blew meat powder into its mouth and recorded accurately the amount it salivated. Then he associated the sound of a bell with the meat powder and repeated this procedure many, many times at successive intervals. The bell, of course, did not at first elicit salivation, but after repeated pairings with meat, it came to so. In describing this experiment, *Pavlov* called the salivation to the bell a *conditioned reflex* (CR), the bell a *conditioned stimulus* (CS), the salivation to the meat an *unconditioned reflex* (UCR), and the meat itself an *unconditioned stimulus* (UCS). This same experiment has been repeated many times on different animals and with many different stimuli and responses. For example, the UCR may be a flexion of a leg in response to electric shock to the foot (UCS), and if this reflex is paired with the sound of a metronome (CS), that signal will eventually cause a leg flexion (CR).

*Fig. 4.12. Pavolov's testing apparatus.*

Typically, the CR is very similar to the UCR, but it is never completely identical to it. Thus, the best way to describe classical conditioning is as *a process in which a previously neutral stimulus (CS=bell) is enabled to elicit a response (CR=salivation) that it never elicited before training*. In his experiments, *Pavlov* found that the time relation between the CS and the UCS was critical. If the UCS preceded the CS. There was very little, if any, conditioning, and if the UCS by more than about a second, conditioning became more and more difficult to establish instrumental conditioning in which the sensory stimuli involved become more numerous and more complex and in which the animal is given a greater and greater measure of choice in the stimuli it uses.

In general, as we go from the simple to the complex learning tasks, learning becomes more difficult and more easily disrupted by brain injury. With these points in mind, it will be interesting to compare the learning ability of different animals that represent different levels of the phylogenetic scale and thus different levels of development of the nervous system.

## Operant Conditioning or Instrumental Learning

*Operant conditioning* or as it is sometimes called, *instrumental learning,* involves a wide variety of procedures. In each instance the animal learns to associate its behaviour with the consequences of that behaviour–that is, the sequence of events is dependent upon the behaviour of the animal. Usually some type of reward or punishment is involved. The task performed by the animal may be relatively simple, as in rat trained to press a lever in a skinner box. Pressing in this example is rewarded with food or water. Another example is when the animal has learned to run a maze to a goal box to receive the reinforcement. In nature a weasel may learn to associate the odor of mice with locating and catching a meal.

### Mechanisms of Learning

It is easier to describe the learning abilities of animals than it is to understand the mechanisms are known. Completing the picture represents one of the greatest challenges to contemporary biology. There is good evidence that the mechanisms by which new information is first stored in the brain are different from mechanisms of longer-term storage. Brain concussions produce *amnesia* (forgetting) for events immediately prior to the injury, which is why people usually cannot recall what happened to them just before accidents. Short term memory may be based on continuously circulating nervous impulses. A rat or a hamster, just having learned a simple maze, such as a T-shaped runway in which the correct solution is to make a right turn at the junction does not remember having learned the maze if it is given an electric shock to its brain within 5 minute of its training. Slight amnesia effects are still detectable if the shock is given up to an hour after the learning experience.

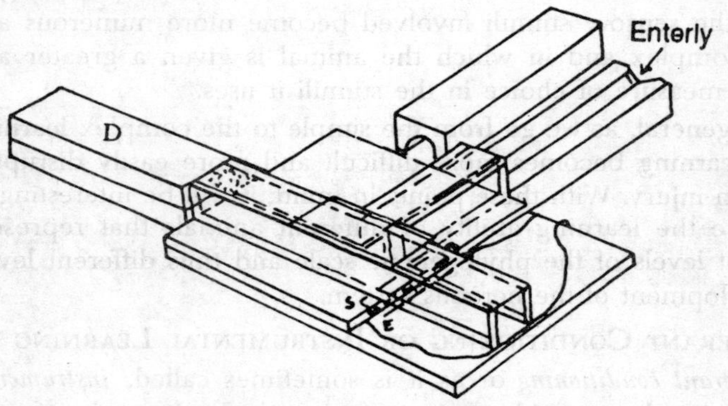


*Fig. 4.13. A T-maze used by Yerkes for training earthworms to turn to the right to enter a dark, moist chamber (D) and to avoid electric shock at E on the left.*

In contrast, long-term memory is undisturbed by convulsions, electric shock, concussion, or deep anaesthesia. For many decades investigators have been trying to determine whether specific memories are stored a particular cells or groups of cells in the brain. Experiments have involved stimulating specific parts of the brain to determine what behaviour is evoked and maling lesions (cuts) in the brain to determine which memories are destroyed by particular lesions. Extensive work with rats reveals that limited regions of the brain are essential for retention of some memories but that particular cells or small groups of cells are not responsible for specific memories. The search for the sites of specific memories a more easily carried out with simpler animals that have fewer cells in their nervous systems. An especially favourable animal has been the marine gastropod mollusc *Aplysia*, commonly known as the sea hare.

As in most molluscs, the nervous systems of *Aplysia* consists of a number of distinct ganglia that innervate different body structures. The cerebral ganglia innervate the anterior tentacles, mouth, eye and rhinophores. The buccal ganglia innervate the pharynx, salivary glands, esophagus, crop, gizzard and muscles that control the protraction and retraction of the odontophore. The pedal ganglia supply the foot, parapodia, head, caudal part of the penis, penis sheath, and penis retractor muscles. The abdominal ganglion controls a variety of respiratory, reproductive, circulatory and excretory functions and also controls movement of the external organs of the mantle. *Aplysia* is not a spectacular learner, but it does habituate to a number of stimuli.

Habituation of the defensive withdrawal of the siphon and gill under control of the abdominal ganglion. Electric shocks normally cause withdrawal of both organs, but if 10 stimuli are given at short (30-second intervals, the animal stops withdrawing its siphon and gills and does not resume doing so far several hours. Four such training sessions, each separated by approximately 1½ hours, produce habituation that lasts up to 3 weeks. Several neural changes are involved in this learning. One is a reduction in transmission of messages from the receptors in the siphon skin and the motor neurons that cause withdrawal. *Eric Kandel* and his associates at Columbia University have shown that just a few neurons decrease the amount of neurotransmitters they release. Conversely, when withdrawal is enhanced by following shocks with other negative reinforcement, a few neutrons release an amine that acts on the terminals of sensory neurons to increase the release of transmitters. Although withdrawal of both organs involves a number of muscles, changes in just a few neurons (the memory traces) are needed for the learning (the modified behaviour) to occur.

## Phylogeny of Learning

We cannot retrace the steps in the evolution of learning, but a brief survey of the learning capacities and capabilities of organisms from various animal phyla should provide some insight into the possible generality of any laws of learning and a sampling of the differences in learning ability across the various phyla. Four cautions should be noted. First, there has been a disproportionate examination of learning across the various phyla. For more studies have been conducted on learning processes in invertebrates than in invertebrates, and even within the vertebrates much of the attention has focused on mammals. Second, there are valid ongoing disagreements among scientists regarding what constitute proper criteria for demonstrating various types of learning and concerning the evaluation and interpretation of learning and concerning the evolution and interpretation of research results. Third exploring the physiological and behavioural aspects of learning in many animals will require more refined techniques and objective, bias-free methods. Fourth, we measure performance and not a genetically determined ability or mechanisms. Lots of factors (notably, test design) can influence performance. Additional problems regarding comparative learning studies will be discussed in two subsequent sections of this chapter. With these difficulties in mind we can now survey the known information regarding learning in a number of animal phyla.

**Protozoa and Coelenterata**

Whether protozoa are capable of learning is still a debated question. It has been reported, for example, that *Amoeba* and *Paramecia* are capable of habituation to noxious sensory stimulus such as strong light or mechanical shock, for upon repeated stimulations their responses grow weaker and weaker until in some cases they become totally unresponsive. However, two criticisms of this work have been offered. One is that none of these experiments has satisfactorily demonstrated that such diminished responses lasted long enough to be anything more than adaptations to sensory stimuli. The other is simply that the noxious stimulation may have temporarily injured the organisms so that they were made capable of responding with each successive stimulation. To get around these objections, attempts have been made to demonstrate some form of associative conditioning or learning. But each claim has again been met with cogent criticisms.

One experiment will serve to illustrate the point. First a sterile platinum wire was lowered into the center of a dish of *Paramecia.* There was no special reaction to it. Next the wire was "baited" with bacteria, and the *Paramecia* responded by congregating around the wire, clinging to it and feeding. Then, after many such presentations of the "baited" wire, it was sterilized and dipped into the same spot, and the *Paramecia* congregated around it and clung to it. This was claimed to be a well-controlled demonstration of learning of a new response to the sterile wire. A simple control, however, demonstrated that the training was unnecessary. In this control experiment, bacteria were dropped into the dish and the *Paramecia* congregated and fed. Then the sterile platinum wire was lowered for the first time into the same spot, and the *Paramecia* clung to it.

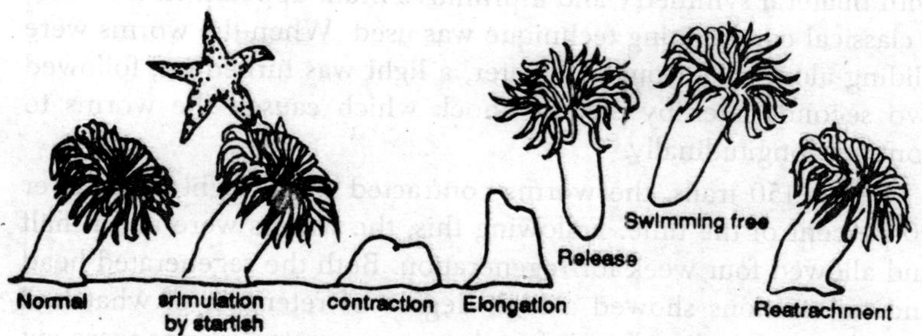


*Fig. 4.14. Interaction between starfish and sea anemone.*

Careful investigation showed that the congregation around the spot the wire contacted was due to the residue of bacteria bait. The increased clinging to the wire resulted from the increased acidity the bacteria contributed to the medium, since further controls showed that clinging in *Paramecia* was a function of acidity. For coelenterates there is also ample evidence for the habituation response, but little evidence for association learning. In the early years of this century investigators generally assumed that coelenterates exhibited only certain involuntary, stereotyped reflexes.

We now know that there are endogenous neural rhythms in many coelenterates and that, rather than being totally "passive," many coelenterates are capable of spontaneous active responses in the course of interacting with their environment (*Rushforth* 1973). One possible demonstration of a form of association learning (*Ross* 1965) is based on the fact that sea anemones (genus *Stomphia*) respond to chemostimulation from certain starfish (e.g., *Dermasterias imbicata*) by stretching their bodies, detaching from the substrate, and "swimming" away. The animal soon lands and eventually reattaches to the substrate. The chemostimulation is paired with gentle pressure applied near the base of the anemone. With repeated trials the application of the pressure stimulus alone leads to some reduction in the "swimming" response; *Ross* labelled this as conditioned inhibition. However, the stimulus may have induced some related effect resulting in the anemone movement.

## Platyhelminthes

Considerable publicity and controversy have surrounded the research conducted on learning to planarians. These flatworms are interesting because it is the platyhelminthes that a nervous system with bilateral symmetry and a primitive brain appears. In this case, a classical conditioning technique was used. When the worms were gliding along in a trough of water, a light was turned on, followed two seconds later by electric shock which caused the worms to contract longitudinally.

After 150 trails, the worms contracted to the light alone over 90 percent of the time. Following this, the worms were cut in half and allowed four week for regeneration. Both the regenerated head and tail sections showed a high degree of retention of what had been learned earlier. Even after these regenerated worms were cut and a second period of regeneration occurred, there was retention of the conditional response. Apparently, in this case, too, learning is not confined to the anterior portion of the nervous system, but has its effect throughout its extent.

**Annelida**

The first unequivocal evidence for learning is found at the level of the worms, where a bilaterally symmetrical, synaptic nervous system is already developed. Here we see clear instances of habituation and associative learning. In one experiment, earthworms were trained to go to one arm of a T-maze leading to a dark, moist chamber and to avoid the other arm which led to electric shock and irritating salt solution. In this simple learning of a position habit, it required about 200 trails on the average to reach a criterion of 90 percent correct responses. Interestingly enough, the worms were able to retain what they had learned after removal of the first five body segments containing the cephalic ganglion, and untrained worms were able to learn after removal of the head ganglion. Apparently, the neural changes involved in learning can take place in the ganglia of the lower body segments.

**Mollusca**

Among the molluscs, some evidence for the learning of a simple T-maze by snails has been demonstrated, but the experiment was not highly successful in that not all snails learned and those that did rarely reached high levels of consistency in their performance. Much better learning has been demonstrated in the octopus, which has a well developed eye and a rather elaborate brain. In these studies, for example, it was shown that the octopus could readily learn to discriminate the presence of a white card signifying electric shock. At first, the octopus was trained to come forward from its

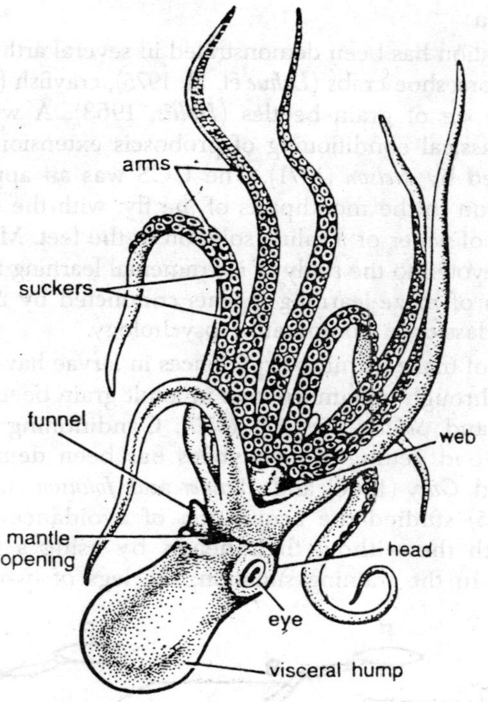


*Fig. 4.15. Octopus.*

rocky nest to seize a crab that was lowered into the far end of the aquarium by a thread. Then on half the trails, a white card was lowered with the crab, and in these trials, the octopus was driven away from the crab by strong electric shock. After 12 trails, the octopus began to inhibit its approach to the card and by 24 trials consistently remained in its nest when the card was present and consistently came out to feed when the crab was presented alone. The octopus can also learn more difficult problems where two different cards were used, a small one associated with feeding and a large one associated with electric shock. Also, discriminations of various shapes have been demonstrated in a similar manner. It is interesting, however, that the octopus was rather poor in problems requiring it to "detour" around a barrier. Thus, when a glass partition was lowered between the crab and an octopus, the octopus persisted in swimming straight into the glass and failed to swim around it through an open space.

## Arthropoda

Habituation has been demonstrated in several arthropod species including horseshoe crabs (*Lahue* et. al. 1975), crayfish (*Krasne,* 1969), and the pupae of grain beetles (*Hollis,* 1963). A well-controlled study of classical conditioning of proboscis extension in blowflies was reported by *Nelson* (1971). The UCS was an application of a sugar solution to the mouthparts of the fly, with the CS being the application of water or a saline solution to the feet. Much attention has been devoted to the study of instrumental learning in arthropods. The studies of maze-learning in ants conducted by *Schneirla* (e.g., 1946) are classics in comparative psychology.

Effects of maze-learning experiences in larvae have been shown to survive through metamorphosis in adult grain beetles (*Borsellino, Pierantoni,* and *Schieti-Cavazza,* 1970). Conditioning of responses related to food getting in honeybees has been demonstrated by *Bermant* and *Gary* (1966) and *Wenner* and *Johnson* (1966). *Horridge* (1962, 1965) studied the acquisition of avoidance responses in insects, with the without their heads, by using a rather clever procedure. In the training situation, the legs of two cockroaches

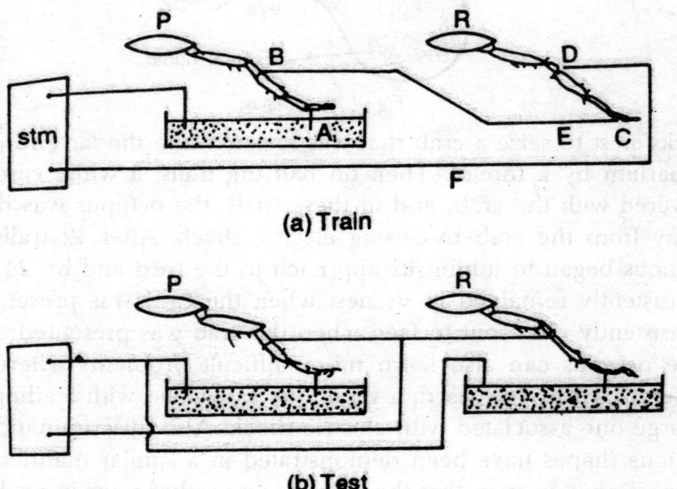


*Fig. 4.16. Experimental arrangement for the conditioning and testing of insect avoidance learning. (a) During training the two animals are arranted so that both animal P and animal R receive shocks when P lower its leg beyond a critical point. (b) In the retest situation, the animal are connected in parallel so that either receives electric shock separately if its leg is lowered.*

are connected in series to an electrical stimulator. The electrical circuit is completed and both animals receive shock whenever animal P lowers its leg into a saline solution, thus completing the electrical circuit. With training, animal P comes to hold up its leg and thus avoid shock. Animal R is a "yoked control" that receives the identical number and temporal patterning of shocks, but they are delivered in such a way as to not be contingent on its behaviour.

Learning is demonstrated as a difference between animals P and R in a test situation when the animals are shocked independently for lowering their legs. Animal P experiences fewer shocks in the test situation than animal R. Both intact and headless locusts and cockroaches have displayed avoidance learning in this situation (*Horridge,* 1962, 1965; *Disterhoft,* 1972). Considerable difficulty has been encountered in demonstrating learning in fruits flies, a species that, in other respects, provides ideal subjects for behavioural studies. Although some learning studies have been quite controversial *Murphey,* 1967; *Yeatman* and *Hirsch,* 1971; *Hay,* 1971; *Bicker* and *Spatz,* (1976), it now appears that fruit flies can learn a discriminated avoidance task based on odor case (*Quinn, Harris* and *Benzen,* 1974).

## Learning in Vertebrates

A number of learning studies have been conducted on species from most vertebrate classes. We know considerably more about the complex processes which underlie these phenomena for this phylum (*Masterton* et. al. 1967); this is particularly true for the mammals. Since the rat and Rhesus monkey have been used in a large number of these studies, an example for each species will provide some flavor of the research and how it is conducted. Rats can be trained in an operant conditioning apparatus to press a lever to receive food. They can also be trained in an apparatus with two levers (designated L for left and R for right) to press the levers alternately, LR or RL, for a food reward. Can rats be conditioned to press the levers in a LLRR or RRLL sequence, called double alternation? Investigators (*Travis-Niedeffer, Niedeffer* and *Davis* 1982) tested this question by conditioning rats first on the single lever task, either R or L, followed by the single alternation task, RL or LR. They then rewarded only double alternation performances.

The rats learned this task at a level exceeding chance expectations, and their performance improved over days. It was necessary, however, in some instances, to permit the rats to give extra responses to the first level before pressing he second lever. Thus LLLRR was rewarded the same as LLRR. When the investigators attempted to condition the rats to a sequence LLRRLLRR or RRLLRRLL (double alternation with a fixed ratio schedule of two repetitions), no rats could successfully perform at better than a chance level. The results are significant because of the new information provided regarding the capacity of the rate to associate a series of responses required to obtain the reward and because previous attempts to condition double-alternation tasks in rats had failed. One particular apparatus, the *Wisconsin General Test Apparatus* (WGTA) has been used in many of the studies of learning behaviour in macaques (*Meyer, Treichler,* and *Meyer* 1965 for the history of this and related techniques). One type of learning problem studied with this apparatus is delayed response problem (*Fletcher* 1965).

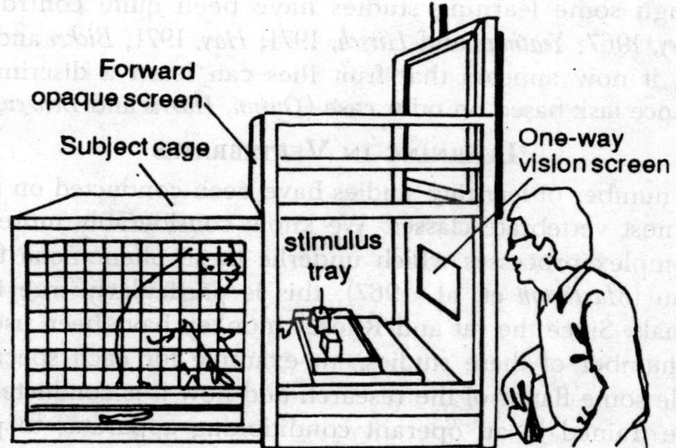


*Fig. 4.17. Wisconsin General Test Apparatus. A early version of the Wisconsin General Test Apparatus (WGTA) used to test aspects of learning in primates and, with some modification of the apparatus, cats and raccoons.*

The basic procedure starts with the test tray out of reach but in full vie of the test animal. Food is placed in one food well, and then two identical objects are placed over the wells. After a prescribed delay the test tray is moved closer to the monkey, which then responds by lifting one object or the other. The four stages in the procedure are the baiting phase, covering phase, delay phase, and response phase. A trail ends when the monkey picks up one of the objects, uncovering either the correct food well, containing a reward, or the empty well. A number of variables can be investigated with this procedure, and other animals, such as cats and raccoons, have been tested with slight modifications of the apparatus. Among the variables that have been manipulated are length of the delay phase, the nature and size of the reward, the nature and the similarity or dissimilarity of the objects used to cover the food wells, and whether the animal is permitted to watch the test tray during the delay phase or, instead, has an opaque screen lowered in front of the tray (*Meyer* and *Harlow* 1952). Among the conclusions are these : Rhesus monkeys are capable of learning basic discriminations in this procedure with delays of up to 30 seconds or more, more food reward leads to performance, and imposition of the opaque screen during the delay phase increases error rates by up to 50 percent. One interesting and striking finding in these studies is that the behaviours and performances of individual monkeys differ markedly.

In general, monkeys that exhibit hyperactivity in the test situation and those that are more easily distracted during the delay phase exhibit lower levels of performance. Clearly, in studies of learning behaviour we must consider the significance of individual differences in performance, regardless of the species being tested, or the task being performed (*Warren* 1973).

## Neural Mechanisms of Learning

If we were to attribute in improvement of learning ability to one thing in the phylogenetic series, it would be the evolution of the central nervous system. It is an article of faith that learning represents some change in the central nervous system, and that memory is the preservation of that change. From this starting point, many investigators have sought an answer to two major questions: (1) where does learning take place in the nervous system, and (2) what is the nature of the change? We shall take up each of these questions separately although it is obvious that they are interrelated. The question of the *locus of learning* has been approached mainly by the technique of experimentally destroying parts of the nervous system.

Most investigators have dealt with the cerebral cortex, since the earliest theories held that it was in this newly evolved part of the brain that mammalian learning occurs. *Pavlov* believed that the cortex is essential for conditioning, but studies have shown that simple conditioning is possible in the dog after its cortex has been removed. Such a decorticate dog often give emotional and generalized responses and is greatly deficient in sensory capacity, but it can be successfully trained in the classical conditioning technique using shock as the UCS. Since total decortication grossly impairs the animal, many investigations of the cortex have involved the destruction of selected parts of the cortex. Thus, in his experiments on rats, *Lashley* just the visual area in the back of the cortex and tested the animals for visual learning and retention. He used a simultaneous discrimination situation in the jumping stand where the animal had to choose the correct one of two doors containing visual stimuli. When he used black and white doors, rats without the visual cortex could learn the discrimination almost normally. If they had learned to discriminate black from white before the lesion of the visual cortex, however, they lost the habit postoperatively and had to learn it all over again. When pattern discrimination was used involving a choice between a triangle and a circle, it turned out that the operated animals could never learn.

Apparently in this case, they lost the capacity for form or detail vision, whereas in the brightness discrimination, capacity was unimpaired and only memory was affected. Actually, however, further studies suggest that even in the brightness discrimination case, it was not memory that was affected, but rather it was a loss of sensory capacity needed to respond to the spatially separated black and white doors. To test this argument, dogs were confronted with a single, large, illuminated panel, shaped like a bowl so as to fill the entire visual field. Then they were conditioned to flex a leg every time the brightness of the field was changed. Here was a brightness discrimination not involving either spatial discrimination or the capacity to discriminate doors, and removal of the visual cortex had no effect on the animal's ability to retain it. So perhaps it was not a defect of memory that Lashley's brain lesions had caused.

Memory and learning ability also proved elusive in Lashely's maze experiments. Here he found that rats were affected in their ability to learn mazes, or retain them, in proportion to the size of the cortical lesions he made in their brains. In other words *Lashley* concluded that the cortex operates on a *mass action principle* in learning and memory so that the large the lesion, the poorer the ability. He also found that it did not mater where the lesion was in the cortex; a lesion of a given size had a given size had a given effect whether it was in the visual area in the back of the brain or the somatic sensory and motor areas in the front of the brain. From this finding, *Lashley* formulated his *principle of equiptentiality,* which says that all parts of the cortex are equal in their contribution to learning and memory.

Again, it seems that these experiments may be as much matter of sensory deficit following cortical lesions as they are the result of defects in learning and memory capacities. Maze learning, we know, is a matter of the rat learning to use many different sensory cues throughout the maze (vision, sound, touch, proprioreception, smell). The more these cues are experimentally eliminated from the animal's sue by destruction of sense organs or by removing stimuli, the worse its performance, regardless of which particular sensory cues are eliminated. Since the rat's cortex is primarily a sensory cortex, it is reasonable to believe that the large the cortical lesion, the more it will impair the use of sensory cues and, therefore, the poorer will be maze performance. Arguing somewhat against this interpretation is an ingenious experiment that *Lashley* performed. He taught blind rats a maze and then removed the visual cortex. Because they showed defects in the retention of the maze habit, he concluded that the visual cortex had nonvisual functions in learning and memory as well as visual functions. Many learning experiments of this type, involving various sensory capacities, have been done on different mammals, and the results have in general been the same. The cortex is not essential for learning or memory. The defects seen after cortical lesions are largely a matter of the sensory defects produced.

One exception to this get real statement is the recent work exploring the temporal cortex of primates. Work with monkeys shows that there are defects in learning touch discriminations following lesions of the posterior borders of the temporal lobe and defects in learning visual discriminations after lesions somewhat more anterior in the temporal lobe. Also, it has been found that human patients with bilateral temporal cortex damage seem to have defects of memory, especially recent memory, as well will see at the end of this chapter. Finally, a most interesting related finding is the fact that electrical stimulation of the temporal cortex of fully awake epileptic patients evokes past memories in a vivid dream-like sequence. However, this may be possibly as much through arousing subcortical structures as through the effect on the cortex itself. Efforts to use the lesion method to explore the role of subcortical structures in learning have not been highly fruitful as yet. Not many such experiments have been done, and what information we have has been largely negative. Two recent studies, however, have provided hopeful leads.

In one, the experiments tried to condition brain-wave responses after making irritative lesions on one side of the brain by implanting aluminium cream. Of all the loci they investigated, placement of the irritative focus just below the temporal cortex in the amygdala and hippocampus was the most effective in impairing learning. In the second experiment monkeys were required to discriminate whether two patterns of tone or of light were the same or different, even though the two patterns might be separated in time by several seconds. In this case, surgical lesion of the region of the amygdala and hippocampus turned out to produce the most marked defects. Such operated monkeys could tell that the two patterns were the same only if one followed immediately after the other. It was as though the lesion made them unable to remember the first pattern over time, for they failed the test when the two patterns were separated by a few seconds. Another approach to the understanding of brain mechanisms underlying learning is through the use of electrical recording methods in which it is possible to trace changes in the electrical activity of many parts of the brain during learning. In these experiments, the animal has electrodes chronically implanted in its brain, and changes in pattern of electrical activity are noted as the animal is trained.

Striking thing here is that the changes take place in many parts of the brain, cortically and subcortically, within the sensory systems and outside of them as well. It may be that the electrical method is so sensitive in recording changes that go on in learning that it cannot separate the important from the unimportant parts of the brain for learning and memory. Or it may be that learning, or different facets of learning, occur in many places of the brain at once, and that therefore, no one part is completely essential for its formation or retention. When we come to the second question, concerning the nature of the change in learning, we find mostly theories and very little facts.

Various mechanisms have been suggested as being responsible for the establishment of new functional connections in the nervous system: (1) the growth of new have pathways, (2) anatomical swilling or sprouting of synaptic terminals, resulting in the facilitation of crossing Grain synapses, (3) a physiological increase in the ease of crossing synaptic connections already established but not functional at the start of training, and (4) biochemical changes such as alterations in the structural arrangement of protein molecules in nerve fibers. Of all these suggestions, one of the most intriguing has been the physiological concept of recurrent nerve circuit in which a loop or circle of connecting neurons is activated such that each neuron in the circle activates the next until the first one, having time to recover, is activated again. Such a loop could theoretically continue firing indefinitely and serve to add facilitation to any synapses in makes outside the loop and thus provides the basis for long-term memory.

At the present time, however, neither this nor any of the other theoretical suggestions have any direct evidence bearing on them. Some insight into the nature of memory mechanisms has been gained by direct, experimental examination of temporal characteristics of the memory process. In one study, rats were trained to run from one compartment to another to avoid an electric shock. They were given one trial a day, and after each trial, they received an electro-convulsive shock through the head. Different groups of rats received the convulsive shock at different times after the learning trail 20 sec,. 1 min, 4 min, 15 min, 1 hr, 4 hr. If the shock came within an hour, there was virtually no learning, but if it came after four, hours, learning was essentially normal. Apparently, memory takes time to "set" in the brain, and this consolidation requires at least an hour, which suggests that memory is a two-part process, consisting of an early phase when memory is vulnerable to convulsive shock and a later phase when it is not. The same kind of conclusion turns up in two other rather different studies.

In one, the octopus was trained, as we mentioned earlier, to discriminate between a crab it could eat and a crab, accompanied by a white card, that it could not approach under penalty of electric shock. If the octopus vertical lobe, an associational region of the brain, was removed, then a curious thing happened. If the trials were spaced more than an hour apart, the animal could not retain enough from trial to trial to improve its performance in normal fashion. If the trials were within fifteen minutes of each other, however, it was able to learn easily. Apparently, lesion of the vertical lobe affected the "permanent" laying down of memories, but did not disturb their "temporary" establishment.

In man, lesions of the temporal cortex on both sides of the brain may result in a similar defect. These patients can learn something simple and retain it for about fifteen minutes to an hour, but after the time, they forget completely and may not even remember having learned. Yet their life-long memories are left undisturbed. Taken together with the animal studies, this finding suggests that memory is a two-part process: (1) an initial, vulnerable, perhaps physiological process lasting fifteen minutes to an hour, and (2) a later, invulnerable, perhaps anatomical process, providing the permanent basis for memory. Many mysteries remain in our quest for the physiological basic of simple learning. We know that learning is a property of is a property of at least all animals possessing a synaptic nervous system. The major question is whether the superior learning of mammals and especially primates is due to the development of superior neural mechanisms for learning and memory or to their greatly increased sensory and motor capacities or both.

The development of superior neural mechanisms must be an important factor, for the capacities for complex learning, problem solution, and reasoning emerge with the evolution of the central nervous system. We have reviewed, in this chapter, the basic facts of animal learning. We began with concept that learning represents an *enduring modification of behaviour* brought about by experience. Then we described various kinds of learning from the simple to the complex: habitation, classical conditioning, instrumental conditioning, and trial-and-error learning.

The essential modification in behaviour in all these cases in the development of some new response to a stimulus that nerve before elicited that response. As to the critical elements of the experience in learning, they appear to be very much the same in all these cases. Or put another way, these various instances of learning including human verbal learning, all seem to obey the same fundamental laws of learning: contiguity, repetition, reinforcement, and for the case of extinction or forgetting interference. When we took up the *phylogenetic development of learning,* we could not find clear-cut and reliable evidence for learning until the level of the worms. This is the point in phylogeny where the bilaterally symmetrical, synaptic nervous system first appears. With the cephalopods and arthropods, which have relatively large, concentrated ganglionic masses in the anterior regions of the nervous system, learning ability is much greater than in the worms.

Finally, with the development of the vertebrate brain, learning capacity develops even further, gradually reaching an asymptote among the simpler mammals. Despite this evidence that relates learning ability to the development of the central nervous system in phylogeny, it has not been easy to discover, with any degree of specificity, the neural basis of learning. The evidence from experimental brain lesions and, more particularly, from studies that record changes in the electrical activity of the brain during learning suggests that learning takes place in many places within the brain at once.

The nature of the neural change in learning has proven elusive, however, despite the fact that many attractive and plausible theories have been proposed. At present, we know that learning or, more particularly, the formation of memories is at least a two-part process. Initially, there is a temporary, perhaps physiological process lasting up to an hour in the mammalian nervous system. Following this and perhaps as a result of it, there a second, more permanent, perhaps anatomical change. Still a third mechanism may subserve the storage of long-standing memories, for it is possible to impair the permanent laying down of new memories by brain lesion without impairing either old, long-standing memories, or the temporary acquisition of new memories.

# 5

# FEEDING BEHAVIOUR

*Feeding* is an important functional system because it provides the source of the energy which is used by animals. The systems includes brain mechanisms, with their many behavioural consequences, which result in food intake, and physiological components which involve, firstly, digestion and secondly, the production, storage and translocation of energy-rich compounds. All of these components, together with the nutrient requirements of the body, must be considered in order to understand the allocation of energy to feeding as a functional system and to each energy-requiring process within the system. The description of anatomical and behavioural adaptations for feeding has been a major activity of biologists for many years. Such descriptions, for a wide variety of animals, are available in scientific textbooks such as *Fretter* and *Graham* or *Welty* as well as in many general natural history books.

Psychologists have used food as a reinforcer in a high proportion of experimental studies on learning. Such work has provided much information about the timing of feeding, the quantities of food taken and the amount of work which will be performed for artificial diet rewards in a laboratory. Information about the brain mechanisms controlling feeding behaviour has come from studies of the effects of the stimulation, ablation or chemical treatment of the brain. Other areas of research on feeding have included studies of human nutrition, the relationships between food intake and production in domestic animals, and the feeding behaviour of insect and rodent pests. Recent ideas in quantitative and behavioural ecology are encouraging interactions between research workers with these different approaches to the study of feeding functional system.

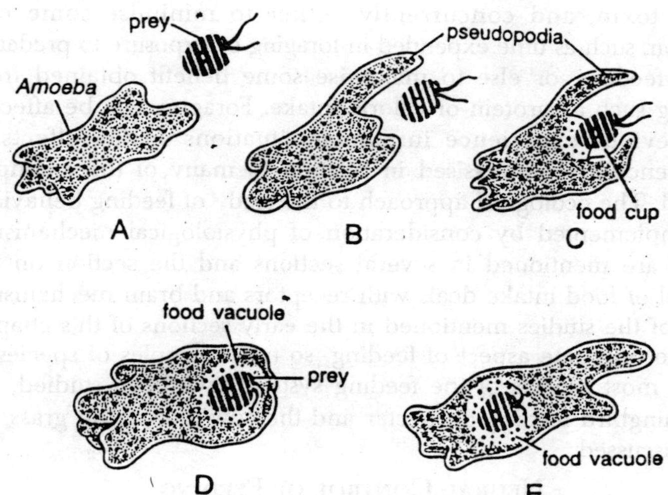


*Fig. 5.1. Amoeba proteus showing formation of food vacuole.*

## Motivational Stage of Feeding

An individual whose motivational state is such that behaviour will be directed towards feeding must take a series of decisions before and during feeding. This series of decisions and the set of mechanisms which makes them possible, is the subject of this chapter. The interrelations between feeding and other functional systems, especially predator avoidance. After a brief discussion of dietary differences and optimality the sections in the chapter refer, in temporal order, to these decisions. Where the word *foraging* is used it refers to the behaviour of animals when they are moving around in such a way that they are likely to encounter and acquire food for themselves or their offspring. Thus, it includes sandpipers finding crustaceans on a beach, sheep grazing in a field, hyaenas hunting wildebeest, and cabbage white butterflies finding and ovipositing on cabbages. Decisions during foraging will depend on potential energy and nutrient returns and on costs. These costs will include those of searching, pursuit, handling and eating; those associated with digestion; those of detoxifying poisonous substances in the food; and deleterious effects of substances in the food.

As *Altmann* and *Wagner* put it 'given the foods that are available to an animal, how much of each should it consume so as to be above its minimum for each nutrient, below the maximum for every toxin, and concurrently, either to minimise some cost function, such as time expended in foraging or exposure to predators while feeding, or else to maximise some benefit obtained from feeding such as protein or caloric intake. Foraging will be affected by previous experience in similar situations so the effects of experience are emphasised in relation to many of the examples quoted. The ecological approach to the study of feeding behaviour is complemented by consideration of physiological mechanisms. These are mentioned in several sections and the section on the control of food intake deals with receptors and brain mechanisms. Most of the studies mentioned in the early sections of this chapter refer to only one aspect of feeding, so two examples of species in which most aspects of the feeding system have been studied, the hummingbird feeding on nectar and the cow grazing on grass are also discussed.

## Neural Control of Feeding

### Hypothalamic Lesions

The hypothalamus, lying in the floor of the third ventricle of the forebrain of mammals, has important functions regulating the coordination of feeding and many other types of behaviour. Damage to the hypothalamus in rats, cats, mice, monkeys, or dogs may have quite different effects on feeding behaviour, depending on the location of the lesion. A rat with a lesion in the region of the ventromedial nucleus of the hypothalamus overeats until it becomes excessively obese; a 250-gram rat may consume food until it weighs more than 600 grams. On the other hand, a rat with damage or anaesthesia localized in the lateral portions of the hypothalamus will stop feeding and, unless it is kept alive by stomach tube feeding, will starve to death.

Electrical stimulation of these areas has the opposite effect– medial stimulation inhibits feeding and lateral stimulation elicits it. Thus, there seem to be separate mechanisms for starting and terminating a bout of feeding. The onset of feeding is associated with activity of the *feeding centres* in the lateral hypothalamus. This activity involves not only the acts of eating and increased responsiveness to food stimuli but also the appetitive phase of locomotion; lesions in the lateral area result in both cessation of feeding and reduction of locomotion. Eating results in consequences which include activation of the 'satiety centres' in the medial hypothalamus. These centres inhibit the feeding centre, and the bout of feeding behaviour is terminated. Later the inhibition is relaxed, the feeding center becomes, active, and another feeding bout ensues.

The reciprocal relationship between the feeding and satiety centres has been confirmed by an elegant experiment using intracranial self-stimulation. In a series of studies made by *Olds*, rats repeatedly stimulated themselves by pressing a lever in a Skinner box. With electrodes in the first position, the area of the lateral hypothalamus including the feeding centres, stimulation was rewarding in a learning situation. With electrodes in the second position, the area of the medial hypothalamus including the satiety centres, stimulation was either neutral or punishing. By preparing rats with several electrode cannulas permitting both electrical stimulation and administration of a local anaesthetic, *Hoebel* and *Teitelbaum* 91962) were able to study directly the effects of one system on the other. They concluded that the medial and lateral centres control self-stimulation in a manner analogous to their control of feeding.

Anaesthetization of the satiety centres resulted both in more feeding and in more self-stimulation in the feeding areas. Stimulation of the medial satiety centre inhibited feeding and also inhibited self-stimulation in the feeding areas. Stimulation of the medial satiety center inhibited feeding and also inhibited self-stimulation of the lateral feeding centres. Previous feeding to satiation also inhibited lateral self-stimulation for periods from 30 minutes to 2 hours. A hungary animal engaged in more lateral self-stimulation, presumably with minimal activity of the satiety centers. Although the hypothalamus clearly plays a major role in regulating food intake in mammals, it is not the only brain centre concerned with this function in mammals. Lesions in the amygdala may also result in overeating in cats, although the hyperphagia is only one-third of that caused by medial hypothalamic lesions in the same species. Other areas of the brain may be more concerned with discrimination and a predilection for certain food items.

## Onset of Feeding

We know from behavioural evidence that the activity of centers controlling the facilitation and inhibition of feeding must be related to the nutritive state of the animal. There is still doubt about how this balance is maintained. Consider the onset of a feeding bout that begins with appetitive behaviour, rather than one that is immediately elicited by external stimuli from food. What activates the feeding center and leads to the onset of locomotion which results in feeding? Early worker thought that stomach contractions, which occur after a period without food, were instrumental in starting feeding behaviour. After long periods of food deprivation they may have an important influence. But various lines of evidence suggest that their rate of contraction is unimportant in normal feeding. A man with a denervated stomach still eats regularly.

Moreover, studies of the effects of various stomach conditions on electrical activity in the hypothalamic centers have shown that stomach contractions are not correlated with activity in the feeding centers, though they are inhibited during activity of the satiety centers. The evidence indicates that some parts of the brain, perhaps the feeding and satiety centers, respond directly to properties of the circulating blood and so instigate feeding behaviour. Alternative or complementary suggestions about the eliciting stimuli, often well supported by evidence, include: a small difference in concentrations of glucose and other metabolites in the arterial and venous circulations, slight drops in the body, and fluctuations in the concentration of a hormone. *Brobeck* concludes that several of these factors and perhaps all will prove to be involved, but that it is not possible at the present time to assess quantitatively the relative significance of each.

**Duration of Feeding**

Once the hungry animal has located food, feeding behaviour is elicited by the external stimuli provided by the food. What factors determine how long feeding will continue before the bout is terminated? Various consequences of feeding are involved. In the first place, the repeated eating and swallowing may itself eventually be self-exhausting as a result either of performance of the motor activity, or of stimulation of receptors in the head and throat. Thus an animal with a fistula which discharges food from the esophagus to the outside rather than to the stomach eats much more than usual. But it does not eat continuously; there are still bouts of feeding. A given quantity of food eaten through the mouth is ore effective in reducing subsequent feeding than the same quantity of food given directly into the stomach.

As a corollary, it has been established that an animal which is stomach-fed through a fistula with as much or more food than it normally eats in a meal still shows an appreciable amount of feeding behaviour. It is therefore clear that oropharyngeal factors participate in terminating about of feeding behaviour, though seldom with a dominant role. Arrival of food in the stomach is critical in the process of satiation. Distension of the stomach wall stimulates stretch receptors which activate the satiety centers in the hypothalamus. Partial inhibition of further feeding can be induced by filling the stomach with inert material. Food is more effective however. Miller has shown this by comparing the effects on subsequent drinking of enriched milk, of enjecting similar quantities of milk or isotonic saline directly into the stomach. Furthermore, an animal given food that is diluted by varying amounts, of inert material will compensate to some extent by eating more. The amount of stomach distension permitted before feeding stops must somehow be related to the nutritive properties of the food; chemoreceptors in the stomach may be responsible. Some further consequences of eating may come rapidly into play, such as the flow of fluids into the digestive tracts accompanying eating, the consequent dehydration of certain tissues, and the release of acetylcholine and other chemical mediators during intestinal absorption.

In general, the assimilation of food is too slow to play a direct role in terminating feeding. Several investigators have pointed out the lack of relationship between the nutritive value of food consumed in a given bout of feeding behaviour and the termination of that bout. A longer-term relationship necessarily exists, resulting in an overall correlation between the nutritive state of the animal and food intake. In this general maintenance of homeostasis, such factors as levels of glucose and other metabolites in the blood presumably come into play .The rise in temperature consequent upon assimilation of a meal may also play a part. In a number of animals warmth and hyperthermia inhibit eating while cold environments encourage it.

## Feeding in Relation to Body Weight

A possible relationship between body weight and food intake is suggested by Teitelbaum's studies of hyperphagic rats. The specific effects of lesions in the satiety centers in causing overeating have already been discussed. In one of the few studies of the detailed temporal patterning of feeding behaviour which have been published, he found that hyperphagic rats given an enriched fluid diet eat larger meals, but that they drink at a normal rate and eat meals with the usual frequency with a solid diet, the increased food consumption in hyperphagia is associated with more frequent and larger meals. Teitelbaum associated this difference with the greater caloric value of the liquid diet. He went on to show that dilution of the liquid diet induced a predicted increase in the frequency of meals in both hyperphagic and normal rats.

Evidently the hyperphagic rat, though it overeats, still regulates the caloric intake of food. Lesions in the satiation centers do not cause completely unregulated feeding behaviour, but something more subtle. Teitelbaum notes that behaviour of a hyperphagic rat changes strikingly when it reaches a certain degree of obesity, and its feeding behaviour comes to resemble that of a normal rat. An obese hyperphagic rate ceases to overeat and consumes only enough to maintain its overweight condition in a steady state. This behaviour suggests that body weight or some correlate of it is

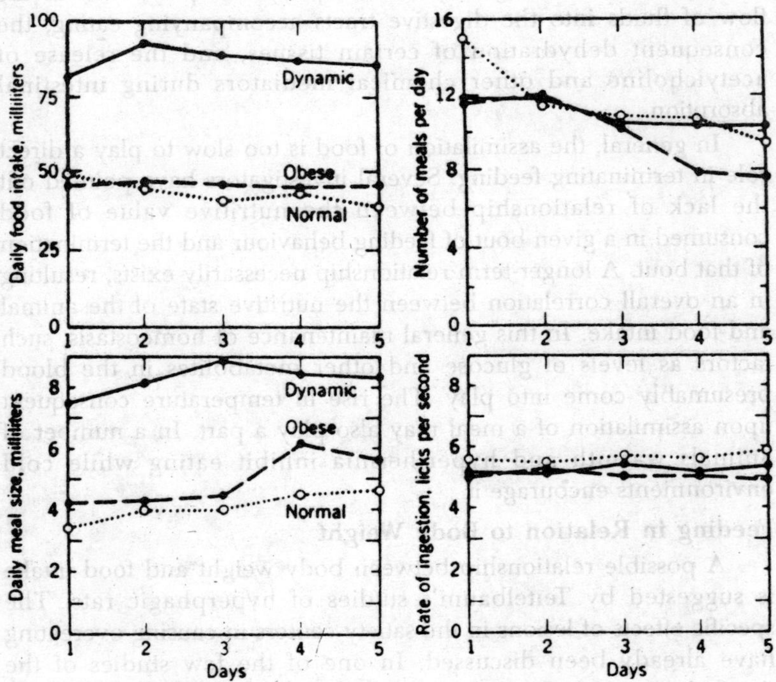


*Fig. 5.2. Several measures of feeding of normal rats, of rats in the process of becoming observes as a result of hypothalamic lesions (dynamic), and of already obese rats. The diet is liquid*

another factor related to the organization of feeding behaviour and that this relationship os disrupted by lesions in the satiety centres. As Teitelbaum puts it, "one might say that the hyperphagic animal overeats to get fat. Once it is fat, it no longer overeats." This notion is reinforced by demonstration that in rats which have been made excessively overweight by other means, lesions cause no hyperphagia. Teitelbaum concludes, with *Kennedy* that the satiety centers may be directly or indirectly responsive to some circulating metabolite related to the state of fat deposits in the animal's body.

**General pattern of Feeding Mechanisms in Mammals**

The tentative picture drawn from these studies of the physiological mechanisms underlying feeding behaviour is as follows. A bout of feeding behaviour starts with locomotion in search of food– the appetitive phase. This phase is triggered by activity in the lateral hypothalamic feeding centers and inactivity in the medial satiety centers. The release of the feeding centers from inhibition by the satiety centers, a response to a wide variety of changes on the body associated with deprivation of food, may be the primary step. Changes in higher brain centers may also release feeding areas from inhibition, allowing for the demonstrable influence of learning on periodic patterns of feeding. After repeated acts of eating and swallowing, followed by arrival of food in the stomach–the consummatory phase–the bout of feeding eventually terminates.

Oropharyngeal factors play an undetermined but definite role. Stomach distension plays a critical role, directly activating the satiety center, which then inhibits the feeding center. Passage of fluids into the gut may have a similar effect. The maximum stomach distension permitted varies with the nature of the food material so that diluted food is taken in greater quantities. The general nutritive state of the animal may also determine, within limits, the degree of stomach distension permitted and thus the duration of the feeding bout. Perhaps the nutritive state directly affects the threshold of responsiveness of the satiety center.

The animal then refrains from feeding for a time, even though it may still be exposed to external stimuli from food materials. The length of the interval before another bout of feeding begins seems to depend on a wide variety of factors including temperature, body weight, and concentration of several metabolites in the blood. In general, it appears that the greater the nutritive deficit, the more rapidly the activity of the satiety center wanes. Feeding resumes more rapidly, and more food is consumed in each bout. Although the duration of feeding bouts and of intervals between bouts varies with the nutritive state of the animal, the rate of actual feeding movements seems to very relatively little under normal conditions.

We are now in a better position to understand the apparent inconsistencies by Miller and *Tugendhat* in the changes of motivation during cycles of feeding and satiation. There is not single, unitary, underlying mechanisms but a multiplicity of mechanisms which are at least in some degree independent of one another. The contrasting results obtained by applying different measures to the same sequence of feeding are no longer paradoxical if we know that many or all of these various physiological factors can influence the animal's behaviour with relative degrees of independence. The sum of their effects is the rhythmical pattern of bouts of feeding separated by periods of other kinds of activity.

This repetitive cycle has no simple basis in exogenous or endogenous influences. The underlying mechanisms involves a subtle integration of effects of stimuli impinging on organisms from without and changes taking place within the organism. External influences do of course loom large, for the behaviour is designed as part of a homeostatic mechanism which must make allowances for variations in the availability, palatability, and nutritive value of different types of food. It is appropriate to consider their contribution in more detail.

## Nature of Diets

Feeding behaviour is often dividend into separate categories according to whether the food material comes from living animals, living plants or dead organisms. Animal morphology is, to a large extent, a result of adaptations for feeding so there are considerable differences between some animal-feeders and some plant feeders, but the above categorisation of feeding behaviour is not always useful. Animals which are suspension feeders may accept animals, plants or organic detritus as food. A warbler may hunt for berries or for insect larvae by methods with similar components. Limpets grazing on rocks eat algal sporelings and barnacle larvae, whilst a nudibranch molluse which is grazing on hydroid colonies will show behavioural similarity in its feeding methods to an insect or mammal grazing on plants. Different digestive mechanisms are required for some components of animal and food but there is much variation within vegetable and animal diets and some plant eaters actually absorb the products of commensal animals (Protozoa) or bacteria.

Although there are of the similarities in the feeding behaviours, digestive processes, etc., of animals eating food of diverse organs, some aspects of feeding are specific to one type of diet. The digestive process in grass-eating ruminants has no parallel amongst species in which the diet is composed of animal materia. Likewise, no plant-eaters needs to chase prey actively so none has need of the adaptations for efficiency of capture possessed by a wolf or a peregrine. Some herbivores are very different in their feeding from some predator but the terms are less useful when considering the many animals with intermediate feeding behaviour and physiology. The most important behaviour which makes feeding possible for parasites in usually host seeking. The behaviour of a parasite looking for a host should perhaps be considered as a form of habitat selection but some of the behavioural mechanisms are similar to those shown by non-parasitic animals when hunting for food.

**Arm Races**

*Daskins* and Krebs have pointed out that the evolution of predator and prey species can be regarded as an arms race. Modifications of genotype continually result in an improvement in the chance that the average individual fox will catch a rabbit or that the average rabbit will be able to evade foxes. A similar arms race exists between herbivores and plants, for plants can produce toxic or unpleasant tasking substances and can bear spines, hairs and other deterrents to would be consumers. Animals can modify their digestive systems and feeding methods to avoid the adverse effects of there plant weapons, Plants and their ability to modify their growth form and their ability to grow in places which are inaccessible to the animals. This arms race is a very slow one which takes many generation and the characteristics of a potential consumer, or a potential food, represent only a few of the many factors which might influence evolutionary changes. The evolutionary competition between lettuces and rabbits is, however, just as real as that between rabbits and foxes. Animals which eat dead matter compete with other organisms which are trying to consume their food, rather than with the food itself.

**Feeding Categories**

A major factor which effects feeding behaviour is the distribution of food. Distribution must be assessed in relation to the size and locomotor ability of the animal. Some animals are surrounded by readily accessible food whereas others may live for long periods without encountering a food item. Before reviewing behavioural methods for finding and acquiring food it will be useful to consider the degree of dietary specialisation or generalisation which is shown by animals. The extent of specialisation which is possible depends upon the abundance and the spatial and temporal distribution of possible food items. Most specialists eat food which is locally or generally abundant but generalists can eat foods which they encounter rarely. Animals which eat only one type of food are called *monophagous.* Those which eat a limited range of foods are referred to as *oligophagous* whilst those which will eat a wide variety of foods are *polyphagous.* Examples of species which are usually monophagous include the giant panda (*Ailuropoda melanoleuca*) which eats bamboo shoots, the snail kite (*Rostramus sociabilits*) which eats the freshwater snail *Pomacea*, the grasshopper (*Gesonula punctifrons*) which eats the water hyacinth (*Eichhornia crassipes*), the chalk-hill blue butterfly caterpillar (*Lysandra coridon*) which eats the leaves of the horse-shoe vetch (*Hippocrepis comosa*), and many parasitic species. Animals which are oligophagous include the koala (*Phascolarctos cinereus*) which eats leaves from about five species of *Eucalyptus* trees, the brent goose (or brant) (*Branta bernicla*) which seldom eats anything other than species of eel grass (*Zostera*), the grasshopper (*Chorthippus parallelus*) which eats grasses, and the larvae of butterflies in the family Pieridae which eat plants of the family Cruciferae with mustard oil in their leaves. Many animals are polyphagous. Examples of species of similar sizes and from the same groups of animals as those mentioned above are bears, rats, man, the jay (*Garrulus glandifarius*), the great bustard (*Otis tarda*), the locust *Schistocerica gregaria*, and the moth *Spodoptera littoralis.*

It seems likely that the ancestral forms of most animal groups were polyphagous. *Emlen* points out that if the members of a species encountered one food much more often and therefore ate it more frequently than others which provided the same benefit in terms of fitness, the average ability to find, ingest and digest that food would probably improve. Such dietary selection would, in a few generations, result in this food being of greater benefit than other foods. Genes which promoted a preference for this food would then spread in the population and the first step towards monophagy would have been taken. The digestion of one food requires a smaller

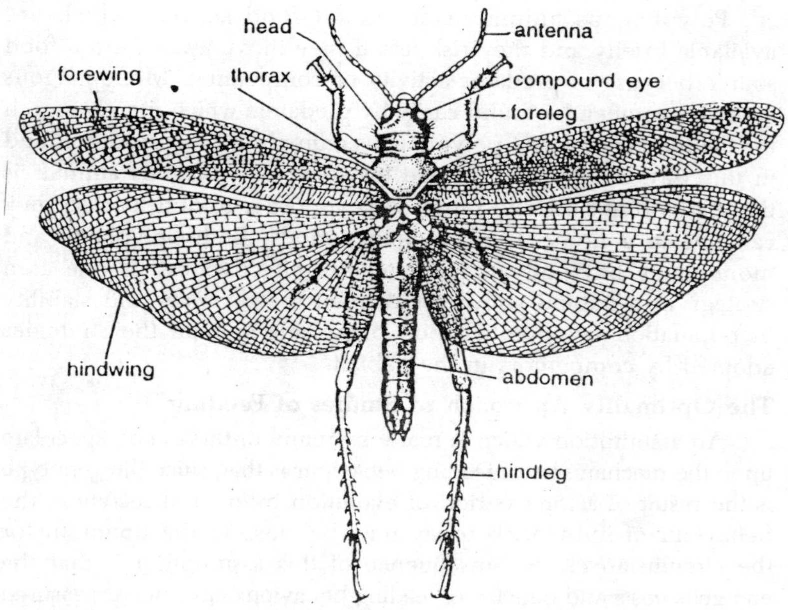


*Fig. 5.3. Locust (Dorsal view).*

range of enzymes than is required by a polyphagous animal. The anatomy and the behavioural repertoire can become more specialised and hence more efficient if only one food is eaten. *Levins* and MacArthur emphasised the monophagy might be advantageous, in terms of the number of young produced, because there is less risk that an inadequate food source would be chosen. A major problem for monophagous animals is what to do when the food source is scarce. Scarcity is a relative term here for it refers to food which cannot be found in quantities adequate for survival by individuals of that species. Some species live through periods during which the food upon which they specialise is absent, by accumulating fat reserves, hibernating, aestivating, encysting, producing long-lived eggs or by hoarding the food when it is abundant. Food is hoarded by squirrels, rats, acorn woodpeckers (*Melanerpes formicivorus*), ants and man. Specialist feeders thus behave in a way which minimises the variance in the availability of their food when they require it. Animals living in unstable habitats cannot to be specialists and thus require a mor complex repertoire of feeding behaviour.

Polyphagous animals can exploit food sources which are available briefly and they risk less if they move away from a food source because of predator activity or competition. Monophagous animals can often be found easily by predators which merely search for the food source but polyphagous animals are less readily found in this way. Another advantage for some polyphagous animals is that they do not have to range so far in order to find food so they can live in a home area for long periods. Both polyphagy and monophagy have advantages and the effectiveness of one as a strategy in a habitat will depend upon the diversity and stability of population of potential food organisms and on the strategies adopted by competitors in the habitat.

## The Optimality Approach to Studies of Feeding

An assumption which is made by many of those who speculate upon the mechanisms of feeding behaviour is that, since the genotype is the result of a long period of evolution by natural selection, the behaviour of individuals today may be close to the optimum for the circumstances. A consequence of this assumption is that the energetic costs and benefits of feeding behaviour have been measured in many studies. The result has been a great improvement in our understanding of feeding mechanisms. The optimal foraging approach advocated by MacArthur and *Pianka* and Emlen has been vindicated by combined studies of feeding ecology, physiology and behaviour.

Any attempt at cost-benefit analysis of feeding must take account of physical constraints, digestion costs and the necessity for acquiring the complete range of essential nutrients as well as the energetic costs of feeding behaviour and the energetic returns from the food. However, it is not sufficient to access optimality in terms of energy obtained in relation to energy expended. The ultimate measure of optimality must refer ideally to the spread of genes in the population. In the absence of altruistic behaviour towards individuals other than offspring, the long-term reproductive potential of the animals under consideration is the factor which is optimised. Statement about optimality of feeding strategies are not useful unless antipredator behaviour, avoidance of other hazards, the regulation of body temperature etc., are also considered. This same argument applies to statements about whether a feeding strategy is evolutionarily stable.

## Finding Food

Members of most animal species have a dispersive stage in their life history so that they need to have the ability to distinguish between areas where food might occur and those where it will not. Once such an area is found, other behavioural mechanisms which increase the chance of encountering food items and make possible their acquisition or rejection, come into operation. The area can then be assessed, the effects of food intake monitored, and further behaviour modified feeding behaviour of some animals includes all of the aspects mentioned in whereas that of others is much simpler. Monophagous animals which spend most of their life eating one individual plant or animal may still have to take decisions about which part of the food organism to eat but their feeding behaviour is less complex than is that of a polyphagous animal which can feed in various habitats.

## Finding an Area

### *Patches and Habitats*

The distribution of the food of most animals is clumped so that the individual must find area os local concentration amidst areas of low concentration or absence. Many people call such an area of local food concentration a patch. Before finding a patch, the individual must find the sort of habitat where sources of its food might occur. The importance of food in determining what can be suitable habitat is emphasised by *Hassell* and *Southwood's* description of a habitat as a collection of patches. For most animals the habitat must have other characteristics as well. Animals often live in a particular area long enough to learn many of its characteristic. Such familiarity is beneficial when evading predators and reducing physical hazards, as well as when seeking food.

Examples of patch-finding include a squirrel finding a nut tree, a mountain sheep finding an area of grass on a rocky mountainside, a goshawk finding flock of pigeons in a field, or a ladybird beetle finding an area on a plant with an aggregation of aphids on it. The amount of food which can be regarded as a patch for a large animals may be a lifetime's supply for a small one. One fruit may sustain a moth caterpillar for the weeks during which its larval development occurs but may be consumed in seconds by a monkey might be, for example, one, several or many ripe fruits on a tree. Species like fruit-eating monkeys often have to travel for sometime between patches whereas leaf-eaters exploit much larger patches and travel less. For example, in a study of the spider monkey (*Ateles geoffroyi*) the animals were shown to spend 28% of daylight in travelling from one forest tree bearing ripe fruit to another. In contrast, the similar-sized black and white colobus monkeys (*Colobus guereza*) ate leaves during 77% of all observed feeding time and travelled for only 5% of the day. These figures are typical of a general relationship for primate species between time spent travelling and proportion of foliage in the diet.

Although there is variation amongst species in the amount of time spent searching for patches, this behaviour is very important for almost all animals. Efficient searching will result in finding better quality patches and hence improve the reproductive potential of the individual.

***Habitat Selection Experiments***

Casual observations of habitat selection have been reported frequently and some detailed experimental studies have been carried out. *Wecker* found that two subspecies of the deer-mouse (*Peromyscus maniculatus*) occurred in two habitats, grassland and mixed oak and hickory woodland. Members of the grassland subspecies kept in an enclosure with grassland on one side and woodland on the other showed a clear preference for the grassland: they are, presumably, better adapted for feeding etc., in grassland. When Wecker reared some mice in the grassland habitat and others in laboratory cages the individuals reared in grassland showed a stronger preference for it. If the mice were reared in the woodland habitat, they also showed a preference, albert weaker, for the grassland habitat.

Observations by *Gibb* showed that coal tits (*Parus ater*), which feed most frequently in coniferous trees, preferred coniferous branches in laboratory experiments whereas blue tits (*Parus caeruleus*), which feed in deciduous trees, preferred deciduous branches. Partridge reared tits of both species in a laboratory environment where they had no experience of any vegetation. When full-grown, the coal tits showed a preference for pine branches whilst the blue tits preferred oak branches. Although the preference was not as marked as that of birds caught in the wild, Partridge's experiments demonstrates that the mechanism which determines the preference can develop in the absence of any experience of the preferred trees. Partridge's conclusion that the preference is 'genetically determined' could be taken to imply that environmental factors are not involved in the development of the mechanisms controlling the behaviour.

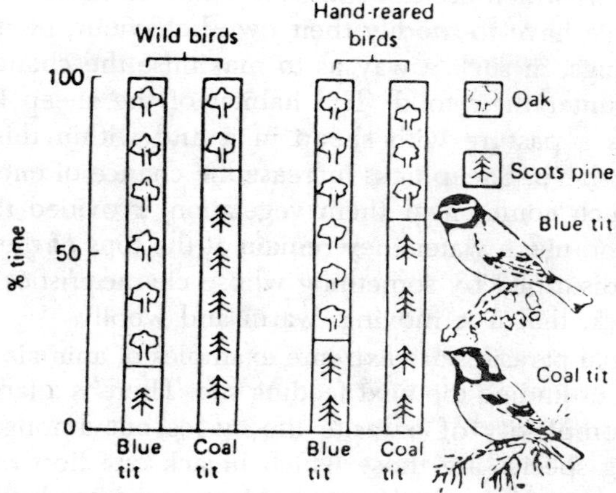


*Fig. 5.4. The proportion of time spent in oak or Scots pine branches by blue tits and coal tits given a simultaneous choice. a, wild birds; b hand-reared birds.*

All behaviour depends on genetic and environmental factors but Partridge's results do show that input which results from contact with the trees is not amongst the environmental factors which affect the development of this preference. Further experiments and observations of the skills associated with feeding in coniferous or deciduous trees showed that the tits preferred the trees for which their skills were best adapted. Studies on a variety of species show that some aspects of habitat preference are modified by experience whilst other are not.

***Host-finding behaviour***

Suitable habitats, and consequently food sources, can be found by moving around and responding to visual and olfactory cues; sometimes at a considerable distance from the habitat. The birds and mammals mentioned above respond to a complex set of sensory cues but an Anopheles mosquito which emerges as an adult after its aquatic early life can find human dwellings by a comparatively simple strategy. If flies across wind until it encounters the odour from human dwellings and then flies up-wing. Such behaviour is similar to that of active aquatic predators, such as sharks, which will swim up concentration gradients towards bodies which are emitting blood and the starfish (*Asterias*) which will move up-stream readily when the body will fluids of their molluscan prey are present in the water.

Animals which are not sufficiently mobile to find the required food source have to modify their own behaviour, or modify their surroundings, in such a way as to maximise the chance that they will encounter their food. The habitat of the sheep tick (Ixodes ricinus) is a pasture with sheep in it and within this habitat, a pitch is a sheep. Sheep ticks increase the chance of encountering a sheep which comes near them vegetation. Provided that they do not lose too much water they remain at the tops of the stems until they are disturbed by something whose characteristics are, ideally for the tick, that it is moving, warm and woolly.

Internal parasites are extreme examples of animals which have problems in finding the next feeding site. There is a large literature on the complexity of parasite life cycles but amongst the most interesting species are those which hi-jack one host or just hitch-hike in order to reach the next. Many parasites have a general debilitatory effect on one host which is then more vulnerable to attack by the parasite's next host. Some parasites, however, modify their environment, that is to say the host, in such a way that the host's behaviour is changed. The acanthocephalan parasite *Polymorphus paradoxus* lives in freshwater amphipod crustacean *Gammarus* and has to get to it next host, the mallard duck (*Anas platyrhynochos*). *Bethel* and Holmes found that, whereas 97% of uninfected *Gammarus* were found in dark areas in an experimental apparatus, only 29% of infected *Gammarus* remained in the dark. The behaviour of the other 80% was altered by the presence of the parasite so that they swam in the light and clung to objects near the water surface. Here, they are much more likely to be eaten by ducks than are individuals which remain in the dark areas near the bottom of a pound or steam. A study of the effects of the eyefluke *Diplostomum spathaceum* on the behaviour of fish showed that heavily parasitised dace and trout spent more time swimming near the surface of the water than did those with few parasites. Here they are more likely to be eaten by gulls which are the nest host of the flukes.

### *Finding patches by observing conspecifics*

The importance of responding to other individuals, often in the same social group, which move in directions which might lead to a food patch is already described. These ideas were initiated by observations of small birds joining flocks which move towards patches, but there are also examples of large birds in the air keeping visual contact with one another so that when one descends, although it might prefer not to share its find, the others fly towards it. This patch-finding technique is used by vultures finding carcases and by seabirds finding concentrations of fish near the surface of the sea. Even if the individuals moving towards a patch are not seen, the patch itself may be rendered much more conspicuous by the presence of an aggregation of conspecifics or individuals of other species which may feed in similar patches. The presence of these individuals is an indicator of environmental quality. The presence of these individuals is an indicator of environmental quality. In experimental studies, Krebs found that great blue herons (*Ardea herodias*) were more likely to land at possible feeding sties at which he had put models of herons.

## Finding Food Items

### *Aggregation in patches*

The efficiency with which patch-finding mechanisms operate in a variety of species is attested to by the aggregation of animals in regions of high food density. Such aggregation occurs in species in which individuals find their way to patch separately as well as in species whose members normally move around in groups. Some examples of animals whose distribution is much affected by that of patches of food. The occurrence of the aggregation is due in part to the use of similar patch-finding mechanisms by many individuals but is accentuated by the greater likelihood that individuals will stay in an area of high food abundance. The advantages of staying in the patch may be counteracted, however, by the disadvantages which result from competition with the other individuals in the patch. The criteria used when deciding whether or not to remain in a patch are discussed later in this section whilst competition for food amongst members of social groups is discussed in chapter 10.

### *Scanning, recognition and reactive distance*

Many animals enhance their chances of finding food by scanning their surroundings. A monkey looking for a fruit, a mantis looking for a fly, or an owl sitting on a perch at night listening for the rustle caused by a mouse, all adjust the head position so that the receptors can detect and locate the food. Such scanning, whether or not accompanied by locomotion, may account for a high proportion of foraging time, for example, in the thrushes observed by Smith. It may be used during patch finding as well as during the detection of items within a patch and is an important factor in food finding by many species.

The *recognition* of food items depends upon the functioning of sense organs, together with sensory analyses in the brain and on the operation of attentional mechanisms. The efficiency with which sensory analysers function is a factor which limits of recognition will also depend upon the background against which the item must be detected. Most research on this topic refers to detection by means of vision or olfaction but some animals use other senses. They may have to modify the environment in some way, e.g., by digging into the ground or by moving vegetation, before they encounter a food item.

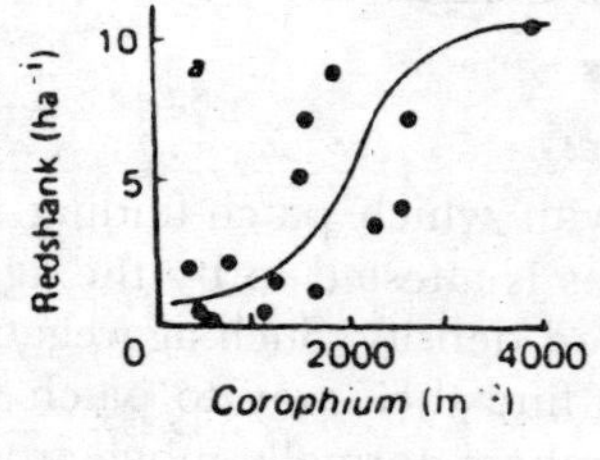


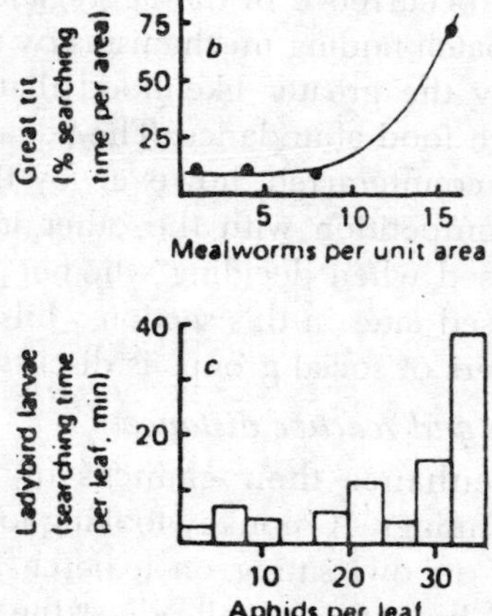


*Fig. 5.5. Foraging strategies often result in the aggregation of animals at food patches. Examples shown here are (a) Tringa totanus (redshank) finding the small crustacean Corophium volutator, (b) Parus major (great tit) finding Tenebrio mollitor (meal worm), (c) Coccinella septempunctata (ladybird) finding Brevicoryne brassicae (aphid).*

*Fig. 5.6. Dace (Leuciscus leuciscus) about to catch the small crustacean Gammarus.*

Precise evidence about the distances at which food is perceived by different species is sparse although we as individuals know the approximate distance at which were recognise mushrooms, berries, shellfish or other items of which we are hunting. Estimates of *reactive distance* have sometimes been made in studies of feeding. *Beukema* recorded the distance at which sticklebacks (*Gasterosteus aculeatus*) turned towards or accelerated towards food items. When the food was the small red worm *Tubifex* the reactive distance was 25 cm. For another small fish, the dace (*Leuciscus leuciscus*) the reactive distance to the amphipod crustacean *Gammarus* varied from 20 to 50 cm according to the numbers of eyeflukes (*Diplostomum*) in the eyes of the fish.

The effect of experience on the reactive distance is considerable in situations where the hunger does not know, initially, the precise characteristics of the prey. When Beukema started to provide sticklebacks with the larvae of the small fly *Drosophila*, they failed to react to the prey at 10 cm distance on most of the first twenty

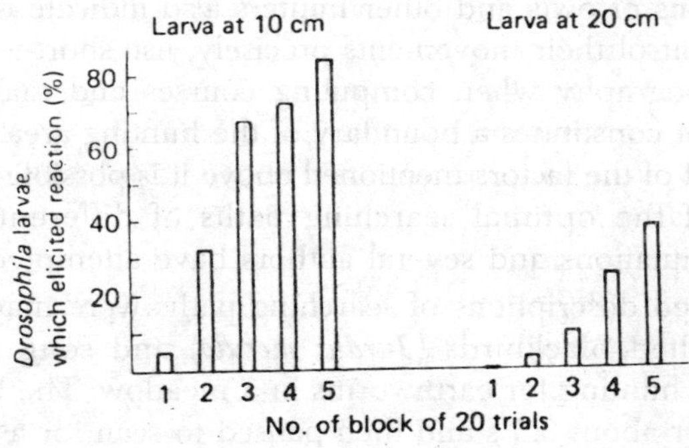


*Fig. 5.7. Sticklebacks do not react often to Drosophila larvae during the first block of 20 trials but the mean reactive distance is less than 10 cm by the third block of trials and is almost 20 cm by the fifth block of trials.*

presentations. By the fifth set of twenty presentations, however, they reacted to almost all larvae at that distance and to some at 20 cm. No stickleback reacted to *Drosophila* larvae at 30 cm but some did react at this distance to the larger, red, wriggling *Tubifex*. The reactive distance to a food item, after extensive experience of feeding on that food alone, is likely to be limited by sensory ability but attentional mechanisms are more likely to determine receive distances in other situations.

**Search Paths**

The movements of individual animals searching for food within a patch have been studied in detail for several species of birds and insects. The search is unlikely to be random for this would involve repeated search in some area and would be inefficient unless the food resource was renewed very rapidly. Ornithologists have long known that owls systematically 'quarter' fields when hunting for small mammals and that feeding flocks of pigeons or starlings move around fields in such a way that they do not recross their own paths. The direction of movement of an owl is, approximately, a straight line except at the edge of the field where the owl swings around so as to return across the field on a path parallel to the first. The distance between successive traverses of a field must depend upon the distance from which the food item can be recognised and attained by the owl. The area around a prey species which is a circle whose radius in this distance, has been called the *danger zone*. A similar term, but one which takes into account the presence of other prey individuals, is the *domain of danger*. These observations of owls and other hunters also indicate that predator species control their movements precisely, use short-term memory of the topography when computing courses and make decisions about what constitutes a boundary of the hunting area. Taking into account all of the factors mentioned above it is possible to formulate models of the optimal searching paths of different hunters in different situations and several authors have attempted to do this.

Detailed descriptions of searching paths were made by Smith who watched blackbirds (*Turdus merula*) and song thrushes (*T. philomelos*) hunting for earthworms in a meadow. The birds moved forward for about 0.5 s and then paused to scan for a mean of 4.8 s before moving again. The mean move length was 34 mm for male blackbird and 450 mm for song thrushes. If no worm was found, the succession of moves often included alternate right and left turns and the beeline direction was approximately straight. If food was found, the thrush often made two or more turns in the same direction and the beeline distance of the twelve moves after capture was shorter than that during the ten moves before capture. In some experiments artificial food, in the form of pastry caterpillars, was provided. Smith found that the search path included more turns and less alternation of turns when thrushes, which had been moving through an area of low food density, encountered an area of high food density. He called this behaviour area concentrated search, a more precise description than 'area restricted search' which other authors have used.

*Drent, Tinbergen* and *Tinbergen* and *Drent* obtained a similar result from observation of starlings hunting leatherjackets, larvae of crane-flies in the family Tipulidae, in grassland. When starlings were hunting for food to take to their nestlings, they were observed to search in comparatively straight lines until leatherjackets were found, then to show area concentrated search. When they had obtained a supply of food they flew off to their nest and then returned to the spot where the high concentration of food had been found. Not only do the birds remember the previous rate at which food had been encountered, when deciding whether or not to adopt area concentrated searching, but they remember localities in open grassland with great precision.

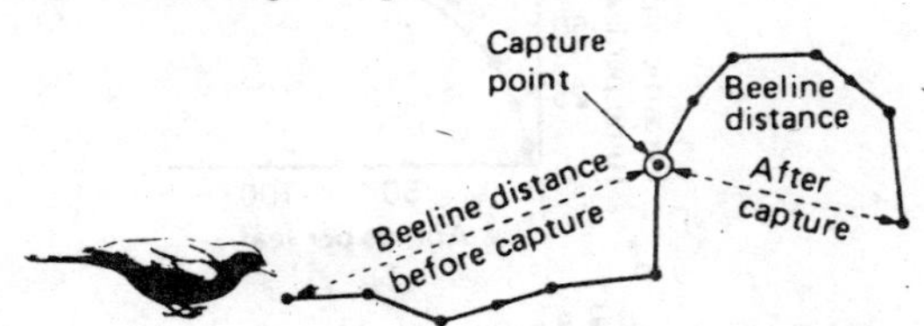


*Fig. 5.8. Movements of a blackbird hunting for worms. The beeline distance during five moves before capture is longer than that after capture.*

The search paths shown by captive oven-birds (*Seiurus aurocapillus*) hunting for mealworms. *Tenebrio* bettle larvae, in a 6m × 6m grassy arena were also affected by present and previous food density. When most of the area of the arena was subdivided into four paths which were provisioned with mealworms at densities of 0.9, 1.8, 3.6 or 7.1m$^2$, the number of visits to a path and length of search path in a patch were much higher in the highest density patch. The logarithms of the number of visits and of the search-path length were proportional to food density. The length of path per visit and the meander ratio (actual path length/beeline distance) were proportional to food density. The overall result of this searching behaviour was that the highest proportion of prey was taken from the highest density patches. When over-birds which had fed in an arena provisioned in this way were returned to it on the following day they immediately started searching in the area which had been most profitable on the previous day. If there was no food there they soon search elsewhere and if there had been two profitable patches they visited both. *Zach* and Falls also provide evidence for the avoidance by ovenbirds of areas which have been

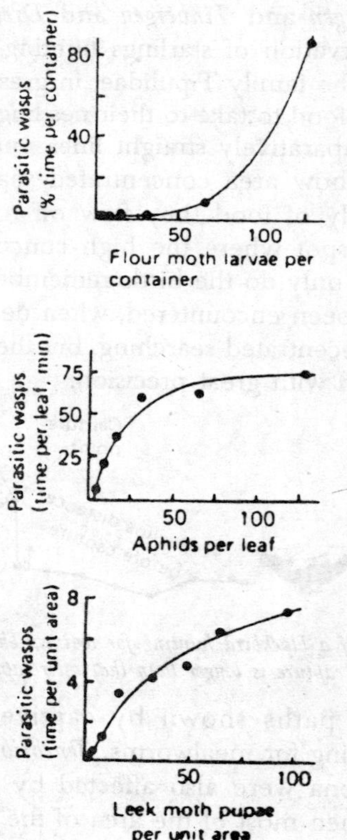


*Fig. 5.9. The search paths of a typical ovenbird Seiurus aurocapillus on day 1, when introduced into an arena with one of the nine squares provisioned with mealworms, and on day 2 when no food was present. The bird returned to the area which had been profitable but left after searching there.*

searched thoroughly, whilst Thomas describes the avoidance by sticklebacks of areas in which possible food items had been found but rejected.

Complex search paths are also found amongst invertebrates, for example the bumble bees *Bombus* studied by *Pyke.* The bees collect nectar and fly from flower to flower in a field on comparatively straight paths but with alternation of left and right turns. In accounting for all aspects of the search paths shown, Pyke concluded that the bumble bee must remember its arrival direction at a flower, its change of direction at a flower, its change of direction at the previous inflorescence and the amount of nectar obtained from the flower just visited.

## Foraging

### *Item size*

The ways in which animals deal with situations where there is only one type of food item have been considered. Alternative food items may vary in size, and hence in energy return when the item is ingested. Just as there are components in feeding behaviour which increase the probability of obtaining more food items per unit of energy expended, optimal foraging theory predicts that if items give a greater net energy return they are more likely to be selected.

Where animals of various species are offered the opportunity to search for and to acquire food items of different sizes, such that the larger items are more energetically rewarding, the proportion of larger items taken is larger than that available in the population, especially at high food densities. Examples include bluegill sunfish (*Lepomis macrochirus*) eating the crustacean Daphnia great tits eating mealworm pieces redshank (*Tinga Totanus*) eating Nereis worms and shore crabs (*Carcinus maenas*) eating mussels (*Mytilus edulis*).

### *Nutrient quality*

If the potential food items which might be encountered differ in quality as well as in size, animals use behavioural mechanisms to maximise the chance that they will ingest items which provide the most energetic and nutrient benefit. In polyphagous species a mixture of different types of food is often needed. The mechanisms used to obtain the food items of the best size and quality often involve taking decisions about searching behaviour and about whether or not to acquire an item once it is located. Sometimes changes in searching behaviour are obviously dependent on previous

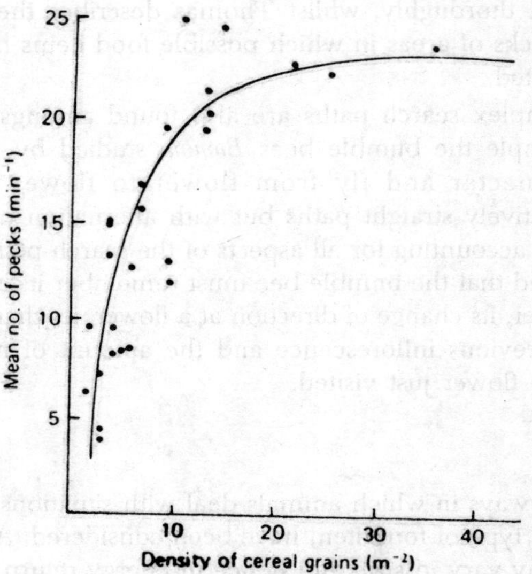


*Fig. 5.10. The rate of pecking at cereal grains by woodpigeons feeding in a field has minimum and maximum values as shown.*

experience. The starlings studied by Tinbergen and Drent (1980) fed their young on a mixture of foods of which the majority (55-80%) were leatherjackets but there was always a proportion of caterpillars of the moth *Cerapterix*, even when leatherjackets were plentiful. A starling which had been collecting leatherjackets from grassland and bringing them back to its nest would fly off in a different direction to a specific area of the salt-marsh where the *Cerapterix* larvae where to be found.

Other examples of observed changes in food searching come from primate studies, for example *Chivers* found that the siamang *Symphalangus syndactylus*, an ape which he observed in Malaysia, would climb to the ends of branches in the early part of the day and eat fruit. At some point in the morning it switched to the central trunk region and ate leaves, thus obtaining a variety of nutrients each day. Much more complex sequences of movements to different feeding sites are observed in species of primates which eat a greater variety of foods. Baboons will move from site to site eating different types of flowers, fruits, grasses, rhizomes, insects, birds and mammals. Many of the sites visited by these and other animals are familiar to the individual so it knows where to search as well as how to search for this type of food. Whilst searching for food in any particular area, however, it is possible that more than one type of food might be encountered. Attempts to relate searching methods to the frequency of occurrence of each of several types of food in a feeding area have been the subject of much research since the studies of L. Tinbergen (1960).

***Item density***

The relationship between the number of food items acquired by an animal and the density of occurrence of these items were described as type 1, type 2 and type 3 functional responses by *Holling*. In his *type 1 functional response* the rate of acquisition is directly proportional to density up to maximum acquisition rate. He describes the occirrseme of such a response in animals which are filter feeders. The other two types of functional response occur as a result of behavioural mechanisms and are explained in this section. An example of the parabolic *type 2 functional response* whilst the sigmoid *type 3 functional response*. In the example the rate of pecking at newly sown cereal grains by woodpigeons (*Columba palumbus*) varied according to grain density, but no pecking occurred at densities of less than $2m^{-2}$ and no further increase in pecking rate occurred at densities of over $30m^2$. *Murton, Isaacson* and *Westwood* suggested that the lower limit was due to the low nutrient return per unit of energy expended at this density whilst the upper limit was a consequence of the time taken to acquire the grain. The importance of this *handling time* was further emphasised by Holling in experiments with human and insect hunters.

Tinbergen recorded the frequency with which small insectivorous birds, especially the great tit (Parus major), brought insects of different species to their young. He related these results to the frequency of occurrence of these insects in the pine-forest

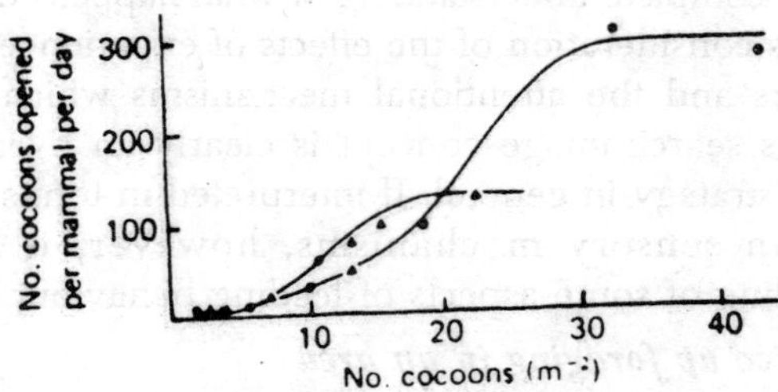


*Fig. 5.11. The relationship between the number of food items (sawfly cocoons) taken by the deer-mouse Peromyscus (circles) and the shrew Sorex (triangles); the density of food items shows a type 3 functional response.*

habitat of the birds. Most rare insects were taken by the birds less often than chance would lead one to expect and common ones more often. This result is very similar to that obtained by *Holling* who studied feeding by small mammals on sawfly larvae in grassland and in laboratory experiments. Most of the food items, in these two sets of experiments, were cryptically coloured and Tinbergen proposed that they were not recognised as food when the rate at which they were encountered was low. Once the encounter rate rose above a certain level the items were taken by the hunger, which they adopted a *specific search image* for this type of food. This idea of modification in perceptual mechanisms has been widely criticised, for example by *Dawkins* and by *Royama* who reported that tits often did not collect the same type of food items on successive visits to the same area.

Despite the existence of data which do not fit Tinbergen's theory, there is no doubt that individual hunters do sometimes show long sequences of taking one type of food when another energetically equivalent food is readily available. Murton conducted a field experiment with wood pigeons feeding on cereal stubble or clover pasture. He provided maize, tic beans, maple peas or green peas all of which had been treated with a-chloralose, which stupefies pigeons. After sometime he collected the pigeons and examined the contents of their crops. When equal amounts of tic beans or maple peas were evenly scattered in a field, one pigeon ate only tic beans whilst others ate only ample pears, and the majority showed a clear preference for one or the other food. Murton points out, however, that due to social facilitation within the flock, his data do not fit with all the ideas produced by Tinbergen: Discussions about methods of obtaining food are based on a variety of casual factors so a complete understanding of what happens during foraging necessitates consideration of the effects of experience, motivational mechanisms and the attentional mechanisms which they modify. Tinbergen's search-image concept is clearly an oversimplification of hunting strategy in general. If interpreted in terms of attentional rather than sensory mechanisms, however, it helps in the understanding of some aspects of feeding behaviour.

***When to give up foraging in an area***

If food is not found in an area the animal must have a mechanism for deciding when to give up searching there. Such a mechanism would also be used after some food had been consumed but no further food was being found. The *giving up time*, during which no food is found, was studied by *Krebs, Ryan* and *Charnov.* The likelihood of giving up should increase as environmental quality declines. Charnov emphasised that the forager must relate intake in the patch to average intake possible elsewhere. It is also necessary for the forager to consider the costs of getting to another patch and the variability of food supplies. This problems is considered further in the section on hummingbird feeding, by Pyke et al., and by Pyke as well as by the other above mentioned authors.

**Eating: Energy, Nutrients and Dangers**

Once a potential food item has been found by an individual, the decision as to whether or not to attempt to acquire it may depend upon a variety of factors. The item will provide benefit in terms of energy and nutrients but the acquisition and digestion of it will involve energetic costs, the time spent on it cannot be spent upon some other item, there could be costs consequent upon resistance to capture by animal prey, there might be costs associated with the physical or chemical characteristic of the food, and there could be costs due to risk of predation or other hazard during the acquisition and digestion of the food. All of these factors, which are detailed below, must be considered at this time, just as they must during food finding behaviour.

**Assessing Energetic and Nutrient Value**

A redshank which has arrived at a point on a mud flat where it can see and reach two holes inhabited by Nereis is likely to choose the larger worm. Such a worm provides more energy, once digested, and is not much more energetically costly to catch. Similarly, a monkey will choose the larger of two fruits and a dog will pick up the larger of two pieces of meat. If the redshank sees a *Nereis* burrow or the burrow of the amphipod crustacean *Corophium* it choses the *Corophium* despite the fact that the net energy return from catching and eating *Corophium* is considerably less than that from *Nereis.* It must be presumed that a *Corophium* diet is more efficient than a Nereis diet as a means of obtaining an adequate balance of energy and nutrients but there is very little direct evidence concerning the mechanisms of such dietary balancing. A similar example is Smith's finding that howler monkeys (*Alouatta palliata*) can assimilate much more energy from fruit than from leaves but they eat both. Again we· assume that they can thereby obtain a balanced diet.

*Fig. 5.12. Redshank (Tringa totanus), feeding on a mud-flat.*

The digestive functioning of animal species has evolved at the same time as have the mechanisms for food preference and the ability to learn about dietary requirements. Monophagous animals can obtain all necessary nutrients from their single food source but polyphagous animals, because of their digestive specialisations, usually need a variety of foods, for example, man and the rat must obtain, respectively, nine and ten amino acids from their food in order to survive. There are mechanisms for ensuring that adequate water and energy-giving food are ingested but the only substances which mammals can detect directly in food and ingest when they are deficient in the body, are water and sodium. There is some evidence for a specific calcium appetite in birds. All other balancing of nutrients appears to occur as a result of feeding preferences which are modified by experience.

Animals can learn about diets during development but must also be able to modify food-selection behaviour according to later experience. Young yellow baboons (*Papio cynocephalus*) will sample food which other baboons are eating and will approach individuals which have food in their mouths and sniff the food. The food choices of the baboons are thus influenced by the choices of the other members of their group and this provides a method of avoiding the necessity to sample harmful foods as well as ensuring that the individual is aware of the existence of enough foods to make a balanced diet possible. It does not, however, result inevitably in the consumption of a diet which is optimal in its composition. There must be much variation amongst individuals in diet, that is to say too individuals may not come to the same decision when confronted with a potential food item.

If an individual is suffering from a dietary deficiency it can modify its behaviour in two ways in order to remedy this situation. Firstly, it can refuse to eat the diet which has resulted in the malaise which it presumably feels. *Rozin* found that rats fed on a thiamine deficient diet refused the diet after a few days, i.e., they treated it like a position and took a new diet preferentially. Secondly, the individual can learn to eat diets which remedy deficiency symptoms. *Garcia*, *Ervin*, *Yorke* and *Koelling* found that preference for diets containing thiamine was enhanced when the diet was consumed by thiamine-deficient rats. The result of such mechanisms is, as Richter demonstrated with rats, that individuals which have a variety of foods available to them, have the ability to compensate for the variation in nutrient needs which results from pregnancy, lactation, thyroidectomy etc.

**Assessing Costs of Acquisition and Digestion**

The energy required to acquire food once it has been found may be considerable and in some species it is very much greater than the energetic costs of searching. A cheetah which sees an antelope, or a baboon which sees a coconut or ostrich's egg, still has much to do before it gets a meal. The oystercatcher (*Haematopus ostralegus*) obtains the flesh of the bivalve mollusc *Scrobicularia plana* by probing down the hole in the mud made by the molluscs siphons, grasping the shell, walking in a circle whilst pulling and then opening the shell valves.

In studies of wolves preying upon ungulates, reviewed by *Mech*, the majority of the animals killed were found to be young, or old, or to have some disability. The most likely explanation for this is that these are the easiest animals to catch. In Royama's studies of feeding by great tits the birds were found to concentrate on slow-moving species of spiders, flies and sawflies. Prolonged chases of active prey are energetically expensive and often may be fruitless. This is something which many young animals learn during their development. The experience gained from previous attempts to acquire food items is a major factor affecting many decisions about whether or not to try to acquire a particular item. There have been very many studies of the improvement of various skills when food is a reinforcer, for example, rats or pigeons on complex schedules in operant conditioning experiments, finches pulling strings and chimpanzees using tools. The selection of leaf rather than stem or dead matter by grazing animal is probably related to costs of chewing and digestion as well as to nutrient content. It is likely that digestion costs are a factor which influences decisions about whether to acquire any particular food item but there is little clear evidence for this. The evolution of the mechanisms which result in food preferences must have been affected by digestion costs so that those items which are preferred are usually readily digestible by the individual. Indeed the digestive mechanisms themselves have evolved so that some animals, such as baboons, require low bulk-high-energy food, whereas others thrive on a high-bulk-how-energy food, e.g.., colobus monkeys.

In a choice situation where two items of equal energetic value and with equal acquisition costs are available, optimal foraging hypotheses predict that the item whose digestive costs are least should be chosen. Even if the energy costs of digestion or acquisition are similar for two items, if one item takes longer to acquire or digest it may be avoided. The importance of time as a resource was emphasised by *Schoener* who pointed out that some animals might be time minimisers rather than energy maximisers in their feeding strategies. An animals might have to wait for a long time before a bivalve mollusc opened widely enough to allow its capture. In this case the animal might decide to leave that molluse and rather than wait, attempt to find other food. If a type of food takes a long time to digest, energy return in a given time might be reduced because of the long period when the gut is full.

## Exogenous Factors

### Food Preferences

All animals are selective in choosing food in their natural habitat, some more so than others. No two species living together at the same time and place at exactly the same staple food. This is the competitive exclusion principle. So feeding selectivity is a subject of major biological importance, not only because it has a bearing upon studies of nutrition, but also because of its relationship to problems of interspecific competition. When animals are given a choice they show preferences. "Cattle exhibit preferences not only for certain plant species but for the same species at different stages of growth, and even for the various parts of an individual plant and for individual plants within a species." The preferences of seed-eating birds, rats and primates are equally clear.

The mechanisms underlying such preferences are not well understood. In seed-eating birds ability to husk a seed quickly and efficiently may determine what seeds are selected. The choice of particular seeds for food correlates well with the size and structure of the bill as in, for example, the Galapagos finches. Experience with different kinds of seeds leads each species to concentrate upon the seeds with the largest and most nutritious kernel which it can efficiently handle. Body size and locomotor ability may play just as important a role as structure of the actual feeding equipment. But within a species a strong element of individuality has been noted by many workers, indicating that a wide latitude in selection of food items is possible.

**External Stimuli and Feeding Behaviour**

What is the role of external stimuli in the selection of food and in control of the structure of the feeding bout that ensues? Some external stimuli tend to stop feeding behaviour, encouraging rejection or even vomiting. Features such as texture, hardness, and shape may playa part, but chemical characteristics, both smell and taste, are especially important.

Cattle reject food with a bitter flavour, and grasses with a high coumarin content are often avoided. Food soiled with feces deters feeding in many species. Among phytophagous insects the selection of host plants is often controlled by the presence or absence of inhibitory and often lethal chemicals.

Visual characters are sometimes equally important in discouraging feeding. The 'warning colouration' of many animals, various conspicuous colours and patterns which repeal predators, are a good example. These features are often coupled with a noxious taste, smell, or texture, or an effective weapon such as a sting, and predators learn to avoid the warning colours by experience. Sometimes harmless species mimic noxious ones and thus avoid being preyed upon. Certain colour patterns, such as the eye spots on the wings of moths, will even deter a naive predator.

The effectiveness of inhibiting stimuli on feeding depends on the state of the animal. It is minimal after long periods of food deprivation and maximal after a bout of feeding. Competition between eliciting stimuli of varying effectiveness may also effect the outcome. Thus, although repulsive stimuli are significant in determining food preferences (e.g., in rats, rejection of certain food objects in favour of other may also occur through a variety of other mechanisms, with availability playing an important role.

Facilitating stimuli are not normally thought of as having inhibiting properties. Yet there is a sense in which such effects can accrue when the stimuli are encountered in a certain temporal pattern either by loss of the power to elicit responses or by more positive inhibitory effects.

External stimuli facilitate feeding behaviour in several ways. An animal may be sensitized to food stimuli in such a way that contact with one stimulus lowers the threshold of responsiveness to others. Olfactory stimuli often have sensitizing effects, and other senses may work this way as well. In gregarious species social stimuli may facilitate feeding. The presence of another chicken causes a hen to eat more grain, and properties of the food, such as the size of the pile, have a similar effect. Other typically gregarious species such as certain schooling fish, rats, sheep, and primates also change their feeding habits and eat more when they are with other members of the same species.

The eliciting function of stimuli is that most commonly investigated by behaviourists and physiologists. External stimuli also orient responses in space, an important fact often overlooked in laboratory studies of feeding behaviour. The range of effective eliciting stimuli is broad in some species and narrow in others. But we still have astonishingly little comparative information on the specificity of stimuli which elicit feeding behaviour. Both quantitative and qualitative properties of eliciting stimuli are important. The concentration of compounds that can be tasted in food, for example, is an important variable. Several experiments have shown that the effectiveness of substances in eliciting drinking rises with increasing concentration in a solution and then declines. These parallels do not imply that the same physiological processes are involved. The rejection of strong solutions of saccharin is probably a response to its bitter taste. The more frequent rejection of strong salt solutions, on the other hand, seems to be at least partly related to the dehydration resulting from its ingestion.

An example of the qualitative properties of external stimuli that are important in sensitizing and eliciting feeding behaviour is found in the larva of the diamond-backed moth, *Plutella maculipennis*, which responds only to the mustard oil glucosides found in certain cruciferous plants. Chemicals play a prominent role as facilitating cues in plant-eating animals but vision can be significant. Aphids, for examples are attracted to leaves from a distance, distinguishing them from blue sky by a preference for light with a wavelength above about 500 mμ. Vision is especially important to predators.

The varying valence of qualitatively different food stimuli in evoking feeding provides a potential tool for exploring the motivation of feeding behaviour. If we consider eliciting stimuli as either strong or weak, their effectiveness in eliciting feeding should vary with changing motivation to food. In the interval following a bout a feeding only the strongest eliciting stimuli will induce feeding. As the interval since the last meal increases, strong stimuli are more likely to elicit feeding and eventually weak stimuli will become effective as well. In the extreme case, after extended food deprivation, even stimuli which would never normally be effective may acquire eliciting properties. In these circumstances the effectiveness of inhibitory stimuli will decrease. Once feeding has taken place the cycle will begin again. Such an experiment might provide an ideal example of rhythmical changes of responsiveness accompanying a cyclic pattern of behaviour.

## Feeding Behaviour of Flies

Files have much the same basic pattern of feeding behaviour as mammals. Given free access to food, the pattern is typically cyclic, with periods of feeding activity alternating with longer intervals without feeding. The combined efforts of a number of investigator have provided a comprehensive picture of the physiological basis of this behaviour which has great intrinsic interest. In addition to the parallels with mammalian feeding behaviour there are certain striking contrasts.

### External Stimuli

The role of external stimuli in triggering feeding behaviour in flies is well worked out. Feeding responses are elicited by a limited class of compounds, primarily certain sugars but also certain proteins. Contact chemoreceptors on the tarsi are stimulated. This sensory input results in extension of the proboscis. Extension brings the chemosensory hairs on the aboral surface of the labellum into contact with the sugar.

In response to this stimulation the labellar lobes open, thus bringing the receptors on the oral surface into contact with the sugar. Stimulation of these receptors, as well as of the labellar hairs, initiates sucking. Feeding is thus initiated and driven by input from oral receptors.

Receptors within the pharynx or esophagus may also be involved. There is some evidence of various groups of chemoreceptors with slightly different functions, perhaps having somewhat different thresholds from one another. In addition to eliciting feeding behaviour, chemical stimuli may also orient the fly to food. Alcohol attracts flies from a distance. Whether chemical stimuli can sensitize flies to eliciting stimuli is uncertain and perhaps unlikely, in view of *Evans* and *Barton Browne's* demonstration that removal of the antennae and palps, which bear the olfactory receptors, does not influence the threshold of responsiveness to beef liver.

**Continuation of the Feeding Bout**

Once feeding has been elicited it continues for a certain time and certain rate. Both the rat and the duration of the bout of feeding are controlled by external stimuli. Rate seems to be a function of the intensity of the eliciting stimulus, increasing with the sugar solution concentration but declining with the strongest solutions of all. In addition to taste, viscosity limits the rate, especially at the highest concentrations of sugar solutions.

The duration of a given feeding bout is a function of the time taken for the contact chemoreceptors to become adapted to the eliciting stimulus. Consummation is thus the simple result of loss of responsiveness to the eliciting stimulus resulting from continued exposure to it. The stronger the stimulus, the longer the bout of feeding. The initial threshold of responsiveness of the animal is an additional variable to consider. Evans and Barton Browne measured feeding bouts lasting on the average from 51 to 133 seconds, depending on the length of time the animal had been deprived of food.

**Feeding Thresholds**

After a bout of feeding, the threshold of responsiveness to eliciting stimuli remains high for a time and then gradually declines to a level at which feeding will occur again. The share of the recovery curve depends on the kind of sugar used. The threshold declines somewhat 20 minutes after the end of the previous bout, probably because the effects of adaptation are waning. Then the threshold rises again to maximum before declining once more.

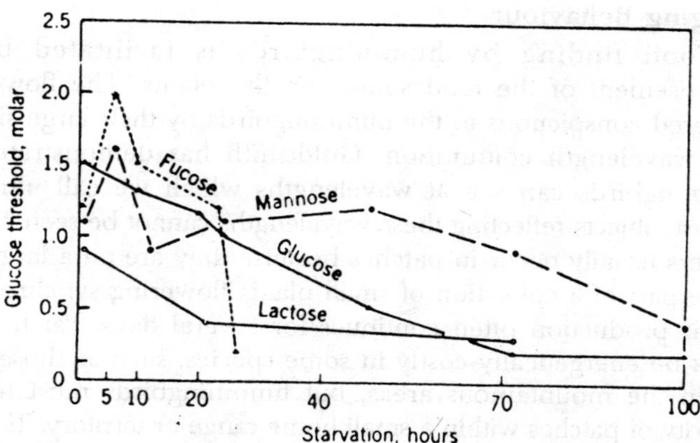


*Fig. 5.13. The minimal glucose concentrations for eliciting feeding in flies after varying periods of starvation, following feeding to satiation with four different sugars.*

This second recovery phase lasts too long to be related to sensory adaptation and suggests a mechanism differing from that involved in the initial threshold decline. This conclusion is confirmed by the discovery that sugars which normally do not elicit feeding and are thus assumed not to be tasted by the tarsal receptors have the same effect of changing the threshold of responsiveness to glucose for a long time. Some consequence of feeding has resulted in a prolonged change in the threshold of responsiveness to eliciting stimuli.

Feeding thresholds are unrelated to the nutritive value of assimilated food. Ingestion of sugars which have no nutritive value for flies, such as fucose, for example, still causes elevation of the feeding threshold. Injection of nutritive sugars into the haemocoel of a starved fly has no effect on the threshold. Furthermore, threshold changes are not related to the timing of changes in the level of normal blood sugar, trehalose. Nor is the simple performance of the motor activity of feeding responsible for satiation, since the detached head of a satiated fly will feed. Experiments involving delicate surgery have localized the origin of inhibition of feeding somewhere in the foregut other than the crop. How the effect is mediated is still uncertain. Evans and *Barton Browne* suggest a possible hormonal link. Chemoreceptors in the wall of the foregut may also play a role.

Locomotory appetitive behaviour for he feeding is closely related to changes in the threshold of the feeding response.

## Foraging Behaviour

Food finding by hummingbirds is facilitated by the advertisement of the food source by the plants. The flowers are rendered conspicuous to the hummingbirds by their large size and long wavelength colouration. Goldsmith has demonstrated that hummingbirds can see at wavelengths which we call ultraviolet because objects reflecting these wavelengths cannot be seen by man. Flowers usually occur in patches because they are on a large plant or are part of a collection of small plants flowering synchronously. Nectar production often continues for several days. Patch finding might be energetically costly in some species, such as those which live in the mountainous areas, but hummingbirds must find the majority of patches within a small home range or territory. Territorial defence in hummingbirds and other nectar feeders is discussed by Carpenter (1978) and Wolf (1978).

The importance of learning the location of patches and the rate of renewal of the nectar supply is greater than that of actually finding the food at most times in the lives of hummingbirds. Foraging behaviour in these birds involves a series of decisions about when to leave a patch and when to revisit a patch. Time spent in a patch of inferior quality could often have been spent in a good quality patch. Before considering such decisions it is necessary to review the information about food intake.

Hummingbirds lick nectar from plant nectaries with their grooved tongues and pass it to the crop. It seems likely that the initiation of feeding depends upon the emptiness of the crop but many other factors must be involved. The size of crop is proportional to the weight of bird. The X-ray studies by these authors show that the food is passed to the rest of the digestive system from the crop, a full crop emptying in 30-40 min in the case of *Eugenes fulgens.* Hummingbirds in the laboratory or in the field usually fly to the food source, be it a feeder or a collection of flowers, ingest enough nectar to half-fill the crop and then return to a perch. This behaviour is repeated with the result that energy is gained at a relatively constant rate during the day.

## Feeding Strategies

Possible feeding strategies which could be adopted by a hummingbird are discussed by de *Benedictis* et al. (1978). The birds could attempt to minimise feeding time but this would involve filling the crop whenever possible, not half filling it as is observed.

The same prediction follows from the strategy of maximising net energy gain per bout. The model of feeding strategy which fitted the observed feeding data best was that in which the rate of net energy gain during the total period from the beginning of one feeding bout to the beginning of the next was maximised. This model took into account the facts that the energetic cost of hovering increases as nectar is ingested, thus increasing body weight, the return flight requires more energy than the outward flight, and the inter-bout period on the perch also involves energetic cost. *Hainsworth* (1978) has extended this model to incorporate overnight storage effects on energetics. The information required for this model includes body mass, duration of light and dark periods, body temperature, ambient night temperature, rate of energy expenditure in relation to body mass and temperature differential, and the energy lost when converting sugars to fat and back again.

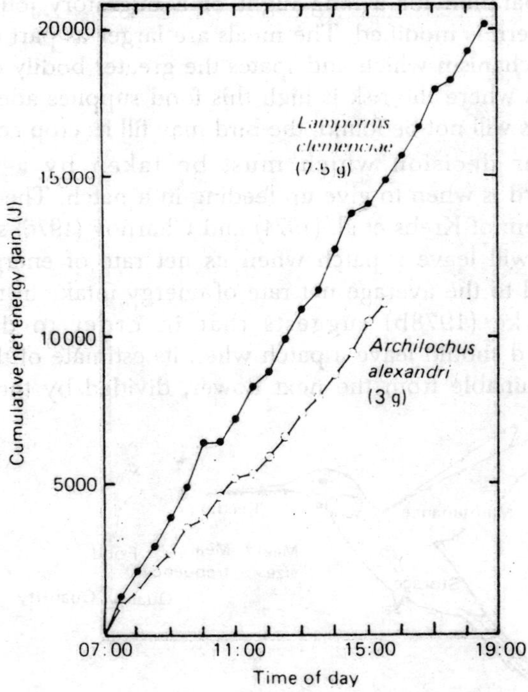


*Fig. 5.14. Hummingbird species of two different sizes showed approximately constant rates of net energy gain during a day. They were fed on 0.5 M sucrose and were kept at 23°C on a 14-h light, 10-h dark regime.*

An estimate of the amount of energy required for storage can this be obtained. This information can be combined with calculation of energy expenditure throughout the feeding period and energy obtained from the food.

Hainsworth's model predicted that maintenance costs during the day would have a major effect on feeding frequency and this was shown to be correct in laboratory studies by reducing temperature, which increased feeding frequency. The model also predicted that increased costs overnight would be remedied by increasing meal size on the following day and this was shown to occur. The function of these mechanisms must be deduced from consideration of the general biology of the hummingbird. The regulatory mechanism in the daytime is that which is the most efficient for maximising the rate of energy gain from the available resources. If the time available for feeding is restricted, such as during preparation for a long night or a migratory journey, the feeding pattern is modified. The meals are larger as part of a feed-forward mechanism which anticipates the greater bodily demands. In situations where the risk is high this food supplies adequate for requirements will not be found, the bird may fill its crop completely.

Another decision which must be taken by a foraging hummingbird is when to give up feeding in a patch. The marginal value theorem of Krebs et al. (1974) and Charnov (1976) states that the animal will leave a patch when its net rate of energy intake has dropped to the average net rate of energy intake in the entire habitat. Pyke (1978b) suggests that in order to do this a hummingbird should leave a patch when its estimate of the nectar volume obtainable from the next flower, divided by the average

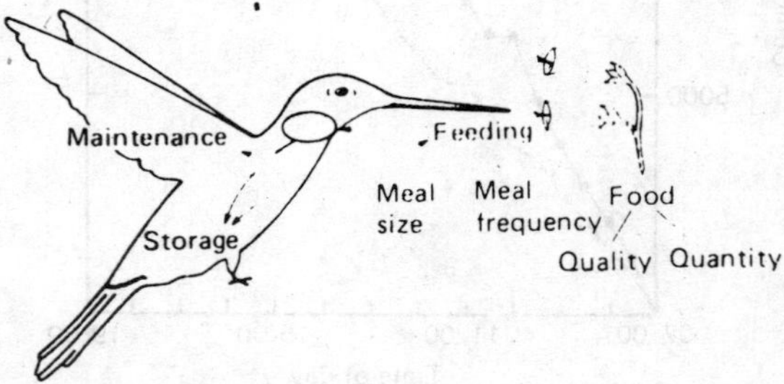


*Fig. 5.15. Hummingbirds consume nectar which is collected from flowers at intervals and stored briefly in the crop before being used for body maintenance or converted to a food reserve, usually fat, for overnight maintenance.*

time required to move to the next flower and remove the nectar, is less than its overall rate of nectar intake in the patch. Observations suggest that hummingbirds use information about the number of flowers in the patch, the number probed and the amount of nectar in the present and the last, or last two, flowers in deciding whether to depart. Although Pyke's model explains his data, it would seem likely that where patches are widely scattered, an estimate of the energy required to move to another patch might be a factor affecting the threshold level of intake at which departure occurs.

Most of the studies of hummingbirds foraging which have been mentioned are of situations where nectar is abundant and feeding behaviour is not much affected by predation risk or by competition for food, nest-sites or mates. Studies of hummingbirds living in mountainous areas where nectar supplies are often sparse and the hummingbird assumes a torpid state during the cold nights, show that vigorous defence of feeding territories occur and community structure is complex. In such competitive situations, foraging behaviour can be inextricably interwoven with other functional systems so that discussions of optimal foraging alone do not go far toward explaining the behaviour which is observed.

**Feeding in Cattle**

The majority of studies of animal feeding in relation to energy utilisation have been carried out on farm animals and some ideas about feeding strategies have arisen as a result of this work. Other ideas have resulted from work on laboratory animals by physiologists or phychologists and work on wild animals by behavioural ecologists (e.g., the study of moose by *Belovsky* (1978) and Belovsky and Jordan (1978). The interchange of such ideas is desirable since many of the same fundamental mechanisms operate the farm, in the laboratory or in the wild. Cattle are ruminants which regurgitate their food from their multicompartmental stomach and chew it during the resting phase after eating is completed. Much of our knowledge of ruminant nutrition comes from indoor feeding studies but field studies are of great importance for world food supplies. Less than 5% of the world's domestic ruminants spent a significant period indoors in 1970 (F.A.O. 1970). The foraging behaviour of cattle and sheep, for example the detailed studies of the late *T.H. Stobs* (e.g. *Chacon* and *Stobbs* 1976) and *G.W. Arnold*, can be related to information about energetic efficiency.

Such studies, therefore, aid in the management of farm animals as well as providing an example of feeding cost-benefit analysis for behavioural ecologists and physiologists.

For many animals, the problems of food-finding limit energy input for ruminants, food-processing takes a long time and is often the limiting factor. The gut usually has food in it so digestion, aided by symbiotic bacteria and protozoans, is a continuous process. The upper limit of the overall rate of processing is determined by the cross-sectional area of the gut and the meal must end once the rumen is full. A consequence of the need for large quantities of food and of slow food processing is increased duration of feeding time and of digestion time. A high proportion of the life of animal, therefore, is devoted to these activities. Wild ungulates, however, also have to find patches of grass or other vegetation upon which to graze or browse. Most wild and domestic ungulates compete with conspecifics for the food which is available. The best way of competing is, usually, to eat as fast as possible when food is found and this behavioural strategy is at least as important to cattle whose grass allowance is carefully calculated, as it is to wild ungulates. Another important objective of the feeding strategy in large generalist herbivores is to *obtain the best mix of nutrients* within a fixed total intake.

## Grazing Decisions

The existence of such behavioural components in the feeding strategy is obvious to the behavioural ecologist but is sometimes overlooked by those concerned with animal husbandry. Cows do not eat randomly all green material which they encounter, as is emphasised in the next section. They have to make a series of decision, when grazing which will differ according to the characteristics of the pasture upon which they are feeding. These decisions include whether to lower the head and bite the herbage, how large a bite to take, at what rate to bite, whether or not to stop biting and chew or otherwise manipulate the grass in the mouth, whether to swing the head to the side and how far, whether to take one or more steps forward and whether to raise the head and carry out some other behaviour. The long-term consequences of such decisions are that total daily grazing duration is 4-14h at bite rates of 40-8- $min^{-1}$ and with bite sizes of 0.05-0.7g. Each day, the food intake may be up to 12.5 kgs dry matter and the cow will ruminate for 4-9h (*Hafez* and *Dyer* 1969, Stobs 1974b, Chacon and Stobbs 1976). In temperate countries cattle feed most in the daytime. Visual food selection is then possible but diurnal. Feeding may be a relic of anti-predator behaviour for animals can feed and keep looking for predator in the daytime but it is safer to remain immobile at night.

## Selection of Grazing

The fact that cattle show feeding preferences is apparent when a grazed field is examined. The sward is less uniform than it was before grazing. Some plant strains and species, usually the more digestible, are preferred to others and leaves are eaten in preference to stems. Some preference studies indicate avoidance of high concentrations of secondary plant chemicals, e.g., tannin, coumarin, alkaloids etc. *Donelly* (1954) found preferences by cattle, among 1206 strains of *Sericea lespedeza*, for low tannin content and fine stems. A variety of other studies has shown clear relationships between alkaloid levels and food preferences of sheep (review by Arnold and *Dudzinski* 1978). Some ruminants, however, have digestive adaptations which allow them to eat plants with alkaloids or other poisons.

Many other studies have related nutritional properties of plants to the extent to which they are eaten by ruminants. There is, as yet no evidence for long-delay learning of nutrient quality by ruminants but the list of deitary requirements of cattle shown in Table (A.R.C. 1965) emphasises the importance of cattle's feeding preferences and the likelihood of occurrence of subtle modifications of diet with experience. It is known that prolonged grazing of grass low in cobalt leads to extreme reduction of intake, which would presumably be associated with movement to a new feeding area it this were possible.

Man limits the choice of food plant for many cattle but plants different stages of development and those contaminated with dung, which is avoided, are still encountered. The composition of pastures is altered by the feeding activities of cattle, by their dung and the fact that they tread heavily on plants as they move around the pasture. Plants which die when grazed or trodden become rarer, whereas those which are avoided become commoner. When the shoot of a grass plant is eaten to within 5 cm of the ground the next shoot is more likely to grow along the ground, so grazing results in a smaller proportion of upright shoots and a larger proportion of horizontal runners and procumbent leaves. Plants near dung pats can survive with a more upright growth form than those in other areas.

Evolutionary changes in anatomy, physiology and behaviour of grazing animals have been a response to the characteristics of grasses but, in response to the grazers, changes in grasses have also occurred. The occurrence of secondary plant chemicals has been mentioned above but the growth from and the mechanical properties of grass must also have changed. The arrangement of fibres in grass in such that cows have to work hard in order to break off the leaves.

**Table 5.1. Daily Dietary Requirements of Cattle**

| *Nutrient* | *Requirement for 400 kg heifer growing at 0.5 kg day$^{-1}$* |
|---|---|
| Calcium | 25.8g |
| phosphorus | 23.7g |
| Magnesium | 6.8g |
| Sodium | 7.5g |
| Potassium | (usually adequate in diets) |
| Chloride | 10.9g |
| Iron | 0.24g$^2$ |
| Copper | 0.8g$^2$ |
| Cobalt | 0.0008g$^2$ |
| Zinc | 0.42g |
| Manganese | 0.32g |
| Iodine | 0.001g |
| Molybdenum | (Trace) |
| B carotene | 0.032 |
| Vitamin D | 100 international units |
| Protein | 210g if low fibre diet |
| Protein | 360g if high fibre diet |

[a]Assumes 8 kgs dry matter intake.

## Input and Output of Energy

Intake by grazing animals can be determined by observing behaviour, recording movements automatically, using an oesophageal fistula, estimating dilution of an indigestible substance, or measuring grass before and after grazing (Stobbs and *Cowper* 1972, *Young* and *Corbett* 1972), *Jameson* (1975).

The potential energy obtained from plant material can be determined by calorimetry. If a known amount of food is consumed by a cow, the metabolisable energy of the food can be calculated by subtracting from the gross energy obtained form it, the energy obtainable from the faeces, urine and fermentation gases. The metabolisable energy provided by pasture is 7-14 kJ $g^1$ dry organic matter. The metabolisable energy may be used for body maintenance, growth or milk production and some is lost as heat. On a high quality diet, 76% of metabolisable energy can be used for body maintenance, assuming the there is no growth or lactation. Additional food above that for maintenance can be used with efficiencies of 62% for growth and 64% for lactation. Interindividual comparisons of energy utilisation must take account of body weight. The logarithm of metabolic rate when fasting divided by the logarithm of the body weight is usually close to 0.75 so the scale factor for body weight which is used in comparative energy studies is weight 0.75 (*Kleiber* 1965).

Energy expenditure can be estimated by measuring respiratory gas exchange. In the laboratory this can be carried out by enclosing the animal in a respiratory chamber and measuring oxygen consumption and carbon-dioxide production (Arms by 1928, Graham 1962, 1965). Alternatively, inspired air flow can be monitored and expired gases can be collected by means of a tube connected to the trachea of the animal (*Blaxter* and *Joyce* 1963, *Colovos* et al. 1970, Young and Corbett 1972a). Neither method is very satisfactory for the metabolic processes may be altered by the treatment, but both of them allow some estimation of the energetic costs of different activity. Indirect estimates of daily energy consumption, by continuous infusion of radioactively labelled sodium bicarbonate and sampling of blood or urine, are similar to those obtained by gas exchange methods. Young (1970) found that young (12-14 months) grazing heifers used 706 KJ $kg^{.75}$ $24th^{-1}$. The results from such studies give values for maintenance energy 20-70% higher for grazers than for animals housed indoors. The total daily energy expenditure estimates are accurate enough to allow the evaluation of particular foods and predictions about requirements. The assessment of the efficiency of feeding strategies, although well worthwhile, is more difficult because results such as those shown in Table 5.2 are variable. External conditions affect costs, e.g., a 10 km $h^{-1}$ wind increased oxygen consumption of sheep by 40%, but sophisticated energy monitoring methods will soon make it possible to access the energetic costs of the various components of feeding strategies.

**Table 5.2. Energetic costs of different activities by cattle and sheep**

| *Activity* | *Cost* | *Reference* |
|---|---|---|
| Cattle | | |
| Lying (after 10 mm +) | 422 KJ $kg^{-0.75}$ $24th^{-1}$ | Colovos et al. 1970 |
| Standing | 290 KJ $kg^{-0.75}$ $24th^{-1}$ | Colovos et al. 1970 |
| Lying–standing | 0.66 KJ $kg^{-075}$ | Colovos et al. 1970 |
| Standing–lying | 0.36 KJ $kg^{-0.75}$ | Colovos et al. 1970 |
| Walking | 2.01 $kg^{-1}m^{-1}$ | Hall and Broady 1934 |
| Sheep | | |
| Grazing | 2.25 J $kg^{-1}h^{-1}$[a] | Graham 1965 |
| Ruminating | 1.00 J $kg^{-1}h$-[1a] | Graham 1965 |

[a] Excess cost over that for lying

## The Control of Grass Intake

The physiological mechanisms controlling food intake by ruminants are reviewed by Arnold (1964) and by *Baile* and *Forbes* (1974). The initiation of grazing by cattle often occurs at particular times during the 24th period and behaviour of individuals is affected by that of others in the social group. It is likely that information from visual, olfactory and gut receptors and from fat depot monitors are factors which affect the start of grazing. The termination of a meal is not normally limited by oropharyngeal factors, for animals whose ingested food is removed via oesophageal fistulas consume much more than they do during normal feeding (*Campling* and *Balch* 1961). When Campling and Balch filled the rumen with ingesta from another animal, however, the meal was terminated early. Water or air-filled balloons had the same effect. Removal of rumen contents extended the meal length unless the intake rate was low because the pasture had already been well grazed (Chacon et al. 1976). As a consequence of such experiments, most authors agree that when the volume of rumen contents reaches a certain level, receptors detect the distension and the meal is terminated (Blaxter 1962, Baile and Forbes 1974).

Since blood glucose levels decline during feeding and injection of glucose into the body does not reduce the rate of feeding unless the dose levels are very high, it is unlikely that glucose levels are of much importance in terminating feeding. Stimulation of the lateral or ventromedial hypothalamus has similar effects in ungulates to those obtained in rats, so these areas are probably involved in the control of feeding, but no gold thioglucose lesions can be produced in ungulates. Baile and Forbes (1974) consider that input from fat-depot monitors may be important in the control of feeding but the evidence is not clear. They also propose that input from monitors of intestinal content would be more efficient in long-term control of eating than would rumen distension monitors since the latter would provide a poorer estimate of true intake.

**Behaviour of Grazing**

The grazing strategies of cattle must include modifications in grazing and ruminating behaviour which depend upon gut contents, upon the quantity and quality and of herbage available now and upon that predicted for future. Domestic cattle have to contend with open range situations, where patches of adequate grazing are widely separated set stocking, where a herd of cattle is kept in a field whose grass renewal rate is sufficient to supply their needs and rotational grazing, where they graze on a paddock or strip of pasture for 1-10 days and are then moved to another. The amount of food available may be barley sufficient for body maintenance or may be enough to allow the production of large quantities of milk. The grass diet may be supplemented with high quality concentrates. These extremes may be encountered within a period of a few days by animals on a rotational grazing system.

When a group of six cows were put into a Setaria pasture in Australia for 14 days by Chacon and Stobbs (1976) they showed initially a very clear preference for leaves but as the total amount of herbage available declined they ate more stem and dead material. The duration of grazing was quite high (over 9th) at the beginning of the 14 days but increased to almost 11th on days 3-6, and then declined. In other studies, grazing time has been shown to increase considerable from lower initial levels as the amount of available herbage declines (*Halley* 1953) Arnold 1964, Stobbs (1974b). In Chacon and Stobbs' experiments the rate of biting increased during the first eight days but the bite size declined throughout the 14 days. The estimated intake dropped a little between the first two and the second two days, despite behavioural compensation and

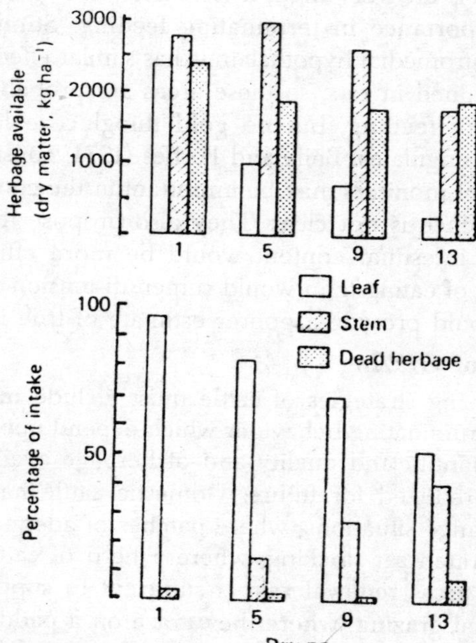


*Fig. 5.16. On day 1 in a Setaria pasture, six cows ate mostly leaf. The amount of leaf available declined but most of their intake was still leaf on the thirteenth day in the pasture. The amount of stem consumed increased during this time but little dead herbage was take.*

dropped considerably after that. These results show that grazing behaviour is modified in a way which is energetically costly but which increases intake when the quality of the food is good. The most costly grazing behaviour is not used, however, when the quality has declined. Such effects of pasture quality on intake are emphasised by *Hodgson, Rodrigues Capriles* and *Fenlon* (1977) and by Baker and Barker (1978). The previous experience of cows in rotations grazing experiments must alter the likelihood that they will expend a lot of energy when the grass has been grazed down. Many cows are quite capable of training farmers to move them to a new paddock when available herbage is low. The cue which they train the farmers to use is the sight and sound of a row of cows standing by the fence and bellowing.

In addition to the rate of biting to the rate of biting and the bite size. Cows can vary manipulatory movements, the amount of walking and the amount lateral head movement which they carry

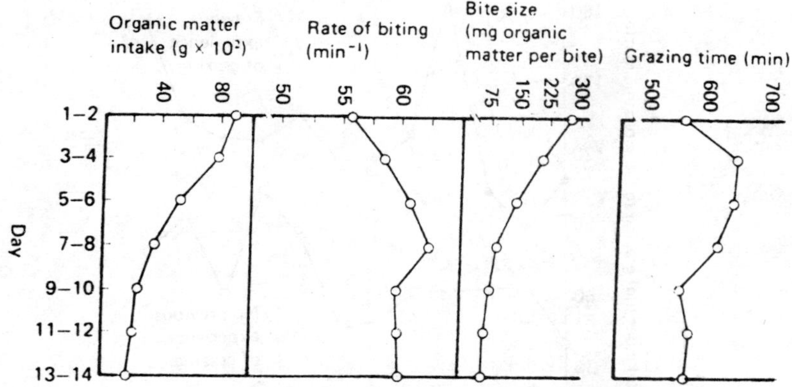


*Fig. 5.17. For the same experiment as that described in 5.16 the changes are shown over the 14 days in grazing time, bite size, rate of biting and total intake.*

out during a feeding bout. A comparison of the same animals grazing on long grass, i.e., mean length of longest part of shoot 30cm, and short grass, i.e., mean length 13 cm, showed that total grazing time was 7.9h per day on short grass but 6.9h on long grass. The mean times spent walking were 56 min and 30 min respectively. When the cows were videotaped whilst grazing, the rate of biting decreased by 8% and the distance walked whilst grazing decreased by 24%. There was an increase of 27% in the number of manipulatory chews. The bite size of cows which had fasted for as little as 16th was about 10% larger than that of cows which had fasted for only 2h (Chacon and Stobbs (1977) but most variation in bite size and rate of biting is related, principally, to the fact that cows will search for leaves and will take them individually in poor pasture.

Another factor which affects grazing behaviour is the presence of faeces on the pasture. The dung from one cow, without decomposition, may cover as much as 200m² in a grazing season and this may affect 10-15% of the pasture required in a season. Cattle will avoid the area close to dung-pats and will preferentially eat clean herbage rather than herbage treated seven weeks earlier with slurry from a cowshed. On slurry-treated pasture cows ate less unless they did not have to eat the grass down to the slurry. If slurry was present they were more likely to stop grazing and walk a few steps and were involved in competitive encounters more frequently than on clean pasture.

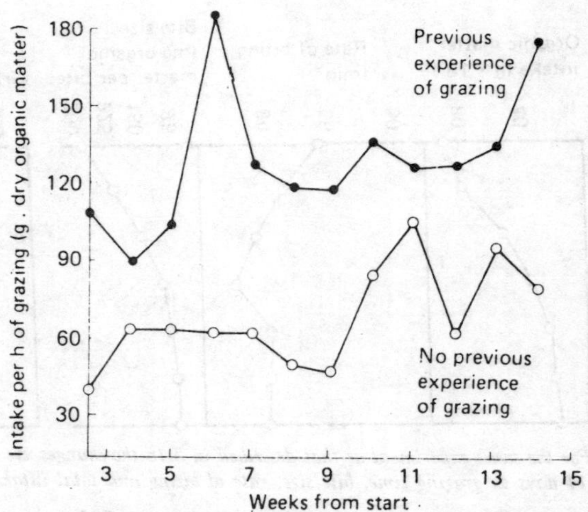


*Fig. 5.18. Three-year-old sheep which have had no experience of grazing (open circles) are less efficient grazers than are sheep which have grazed (filled circles). Their performance improves after ten weeks practice.*

Grazing behaviour is also affected by weather and by disturbance. Grazing ceases in driving rain or if a large predator is heard or smelled. The sight of a dog or the occurrence of any unfamiliar activity on the farm may result in an interruption of grazing behaviour, probably as part of an anti-predator response. Calves are more likely to be disturbed by any particular event, perhaps because cows have had much previous experience of such events which had no adverse effect on them.

The development of grazing in calves involves an increase in grazing time during the first four months as rumen function develops (chambers 195, *Hutchison* et al. 1962). By one year of age, calves graze for an hour a day more than heifers and for 1.6h per day more than 3½ year-old cows (Hodgson and *Wilkinson* 1969). This may be because calves are more selective size or experience of grazing has improved the feeding efficiency of the older animals. *Arnold* and *Maller* (1977) and Arnold and Dudzinski (1978) showed that sheep reared for three years with no grazing experience sheep and also reported that preferences for pastures plants were much affected by previous experience.

## Anti Predator Behaviour

As most species are preyed upon by animals of at least some species, avoidance of predation plays an important role in survival and reproduction.

Some defensive mechanisms operate throughout much of the life of an individual whereas others are used only when a predator is detected or when a predator attacks the individual. *Edmunds* (1947) refers to two types of mechanism, primary and secondary. Following *Kruuk* (1972) Edmunds define primary defence mechanisms as those which operate regardless of whether or not there is a predator in the vicinity. The behavioural components of such mechanisms may occur or before any predator is detected. The mechanisms include (1) hiding in holes; (2) the use of crypsis; (3) mimicking inedible objects; (4) exhibiting a warning of danger to predators; (5) mimicking individuals in category (4); (6) timing activities so as to minimise the chance of detection by a predator; (7) remaining in a situation where any predator attack is likely to be unsuccessful because of possibilities for secondary defence; (8) maintaining vigilance so as to maximise the chance of detecting the advent of a predator. All of these mechanisms decrease the chance that a predator attack will occur. Many of them have a considerable influence upon habitat selection. For some species which are very vulnerable to predation, the avoidance of predators limits the distribution of the animals even more than do factors related to body maintenance and feeding. The selection of a suitable habitat may be accomplished by the recognition of a factor which is not in itself advantageous but which is an indicator of the presence of advantageous features; such as those which make possible feeding or predator avoidance. For example, a bird, such as nightjar, which is camouflaged when amongst dead wood might fly over open downland towards a dark green mass of woodland because suitable hiding places are likely to be found around the edge of the wood.

*Secondary defence mechanisms* are defined by Edmunds ad those which operate during an encounter with a predator. The encounter may just involve the supposed or actual detection of a predator by the prey or it may involve an attack by a predator. Some secondary defences are antirely pesive. The prisim-spined sea urchin *Diadema* or the thick-skinned rhinoceros are adequately protected from attack. No behavioural specialisation is needed. Active forms of secondary defence include (1) exaggerating primary defence, e.g., a camouflaged animal remaining motionless; (2) withdrawal to a safe retreat; (3) flight; (4) use of a display which deters attack; (5) feigning death; (6) behaviour which deflects the attack to the least vulnerable part of the body, to an inanimate object, or to another individual; and (7) retaliation.

Both primary and secondary defence mechanisms are used by most animals and different mechanisms are often used for different predators. In an encounter with a predator, the prey animal may be able to use several alternative anti-predator behaviours. Decisions must be taken about, for example, whether to remain motionless, to flee, to display, or fight. If one method is tried, without successful avoidance of attack, others may be tried. There are many possible strategies which involve alternative courses of action depending upon the characteristics and behaviour of the predator.

## Defensive Responses by Farm Animals to Man

To the ancestors of present-day farm animals, man must have been recognised as a dangerous predator. Despite many generations in domestication, the anti-predator functional system still has considerable effects on the behaviour of farm animals. In some situations, predation by wild animals or dogs is till possible but anti-predator behaviour now is directed against environmental changes who origin in largely unknown to the animals, or against man. The occurrence of unexpected sounds or the arrival of olfactory or visual portents of danger may elicit adrenal and behavioural responses which are the same as those shown by their wild ancestors. The detection of a person who remains at a considerable distance from the animal may elicit no such responses but the close approach which is often necessary on farms must often elicit a defensive response. The fact that people are often treated as predator by farm animals is ignored by many of those who work on farms on who attempt to understand the behaviour of the animals but the competent stockman has learned observation about such behaviour. Good stockmen also know how to minimise the anti-predator behaviour which is directed at them, for if they can do this they can often improve production and enjoy their work more.

The importance of the abilities of the cowman in obtaining the best possible production from a dairy herd was stressed by *Albright* (1971). Milk let-down occurred faster if the cow saw the familiar milker. Yield was higher if the cowman was delioerate in movements, talked quietly to the cows and followed a regular routine with no panic activity. Other effect of a 'good' cowman were that cows entered the milking parlour, and were more approachable in the field. It seems likely that a 'poor' cowman elicits a greater adrenal response and corresponding defensive behaviour so that management of the dairy herd is more difficult and milk production is impaired. Studies of the defensive responses of cows to cowmen and to disturbing aspects of yard and milking parlour design, are helping to improve management and milk production.

The detection of a person in an animal-house may be a traumatic experience for comparatively small animals such as chickens or turkeys. If a person enters a house rapidly and noisly the nearest birds may show a violent escape response. Whether the birds are in separate cages, or in one large aggregation on the floor of the house, a wave of violent escape behaviour, sometimes called 'hysteria', may be transmitted down the house. Such behaviour can be minimised by a quiet knock before entry and by slow, quiet movements while entering and when within the house. The economic importance of quiet slow movements in poultry and other livestock houses is obvious to the turkey farmer who has found the corpses of dead birds under a heap of turnkeys disturbed by the entry of a noisy person. The effects of defensive behaviour and adrenal activity on larger animal may be obvious, for example, if a fleeing cabreaks a leg, but many effect are less obvious. Intermittent high adrenal activity may affect adversely milk or egg production, growth rate an disease resistance as well as ease of handling.

**Warning Colouration and Mimicry**

Some animals protect themselves against being eaten by containing poisonous or sickening substances. Some such animals make their own poisons; others takes them from sources in their environment. The wings of the monarch butterfly, for example, contain powerful heart-stopping poisons called cardiac glycosides. The monarch eats the poisons as a caterpillar, when its food plant is the asclepiad, or 'milkweed', which contains cardiac glycosides. The caterpillar is not harmed by the poisons; it just stores them, and they are then retained by the adult.

For the behavioural problem of defence by poison, we must turn from the prey to the predator. If the defence is to work, the predator must learn not to eat poisonous animals, because natural selection will not favour a trait, by which an animal, after it is dead, makes its attacker sick. If the trait is to evolve, it must ensure survival. The tactic used is to enable predators to learn to recognize sick-making prey, in which case they will avoid them. It is a skill predators will readily learn. When a bird eats a monarch, as an adult or caterpillar, it will be violently sick within minutes, an experience it would learn not to repeat. The bird's problem then is to distinguish sickening from edible prey. Now, the colours of the monarch are bright gold, and ethologies suspect that the bright colouration evolved to make the monarch more memorable to birds. It is therefore called warning colouration. Experiments have shown that predators learn to avoid sickening prey, J.V.Z. Brower, for instance, offered the monarch butterfly as food to the Florida scrub jay. On their first meal of monarch the jays were violently sick; but after only a few trails they had learned not to eat monarchs, though they continued to eat other, tasty food. What has not yet been conclusively demonstrated is that birds learn to avoid brightly coloured sickening prey more quickly than equivalent but duller coloured prey. It is necessary for the theory that they should, for otherwise poisonous prey might just as well be dull as brightly coloured. Few ethologists would doubt, however, that such an experiment would be successful.

The early stages in the evolution of warming colouration pose a paradox. The population of monarchs clearly benefits from teaching each generation of birds not to eat them. They benefit by being eaten less. The problem is that some butterflies have to be eaten to teach the birds to begin with. In evolutionary terms, it is no consolation to the dead butterflies that some others are benefitting from their death. They are dead; they have failed to reproduce; natural selection has worked against them. When warning colouration first appeared in evolution, it would have been a rare, minority characteristic. Those rare, brightly coloured butterflies would have been conspicuous to predators, and therefore eaten. One would expect natural selection to have eliminated the characteristic. How could natural selection favour an increase in its frequency? The question has not been satisfactorily answered, though there are some possible answers. One possible answer is that the family relatives of the eaten butterfly may benefit from its death. Members of the same family tend to resemble each other; if one were warningly coloured, many others would be as well. If one member were killed, the rest of the family would benefit from the lesson taught to the predator.

**Camouflage**

The nochid moth's defence is to seek escape in active flight. The opposite defence is to sit dead still and try to be invisible. Such is the method of camouflage in which a species evolves to resemble its background. Camouflage is of course an adaptation of appearance and colouration, but the most exquisite artistry will be wasted if the animal's behaviour is not suited to the camouflage. The world is a patch-work of different colours: the animal is only camouflaged if it settles in the right place. Consider the European grass hopper *Acrida turrita.* It comes in a green form and a yellow form. In a nature the green form lives in green places and the yellow form in yellow and brown places, with rare exceptions. In a simple experiment, the German ethologist *S. Ergene* gave yellow and green grasshoppers a choice between yellow and green backgrounds. The green grasshoppers fittingly tended to go and settle on the green backgrounds, and the yellow grasshoppers on the yellow.

The North American moth *Melanolophia canadaria* faces a more difficult problem in tining up with its background. It has striped wings and lives on the bark of trees. It must line its stipes up with the lines of the bark if it is to be camouflaged. In an experiment, *T.D. Sargent* allowed striped wings and lives on the bark of trees. It must line its stripes up with the lines of the bark if it is to be camouflaged. In an experiment, T.D. Sargent allowed the moths to sit on cylinders that had regions of vertical stipes and regions of horizontal stripes. If the stripes (which were made of black tape, struck on a white surface) could be felt by the moths, then the mouths usually lined up correctly. When Sargent covered up the stripes and surface with a transparent film, the moths no longer lined up correctly. The moths must be relying on the feel of the surface that they have to line up on. In nature they will be able to feel the stripes of their background, and ensure that they settle in a camouflaged posture.

## Escape

Rapid and agile locomotion provides the best and probably the most common means of escaping from predators. Various species supplement their locomotor escape patterns with displays that function to distract or startle a potential predator. Still others may adopt a state of tonic immobility ("play possum") as a means of reducing the likelihood of attack.

## Social Antipredator Behaviour

### *Confusion Effect*

One of the main benefits of sociality is though to be the protection gained, as stated in the adage, 'safety in number'. Living in flocks herds, or schools increases the chance of spotting predators. When approached or attacked by predators, groups usually become more compact. Fish schools, however, usually form a vacuole around a predator. Tightly packed prey make a difficult target for the predator, which must select an individual toward which it can aim its attack. The predator tries to isolate an individual from the group; any individual that strays from the group or is in any way different from the rest is likely to be attacked. For example, when experimenters presented hawks with set of ten mice, the hawks preferred the oddly coloured mouse *Mueller* (1971).

A primary means of protection is the confusion effect, described by Miller (1922). A variety of predators that attack swarming prey exhibit a lower prey capture rate, a longer period of hesitation, and more irrelevant behaviour than when they attack solitary prey. Factors that enhance the confusion effect the swarm number, swarm density, and the uniformity in appearance of swarm members. The mechanism producing the confusion effect is not understood, but a prenator seems to be more easily frightened when it feeds in a dense swarm of prey; perhaps the predators finds it harder to detect other animals that might prey on it. In laboratory experiments with three-spined sticklebacks (Gasterosteus aculeatus) feeding on water fleas (*Daphnia*), fish with experience in feeding on dense swarms fed more efficiently when tested with dense swarms than did those experienced only in feeding on low-density populations.

***Detection***

We might expect to find groups of mixed species if the combination of different sensory capabilities is advantageous. Baboons and ungulates are frequently together on the African plains. The aboreal habits and keen vision of baboons and the well-developed olfactory systems of ungulates probably increase the distance at which prey can be detected. It is clear that these different species recognize and respond to each other's alarm calls.

Social living increases the likelihood that predators will be detected and allow group members to spend more time in other activities. Individuals in large, dense colonies prairie drops (*Cynomys* spp.) spend less time in alert postures than do those in small colonies. Once predators are detected, alarm calls alert neighbours to danger. *Marler* (1955) has argued that in some birds the alarm call is difficult to locate because it is a high-pitched note in a narrow-frequency range. The alarm-calling ground squirrels studied by *Sherman* (1977) emitted calls that were easily located, and calling squirrels were more likely to be caught than non-callers. Close relatives of the caller were likely to be close by and thus could benefit from this seemly altruistic behaviour.

**Concealment**

Many prey species concerned themselves from predators by remaining in protected locations–such as burrows, crevices, or huts–when predators are present. Alternatively, concealment can be provided by the actual appearance of the animal itself. Cryptic colouration, which functions to enable animals to blend in with the background, can be found in virtually all animal taxa. Many extreme examples are found in insects, which may resembles leaves, twigs, or even bird droppings. Often, a particular behavioural pattern is associated with protective colouration such that the animal orients itself in a particular relationship to its environment, usually remaining motionless.

**Freezing**

Another strategy in response to predators is to freeze. The presence of protective or cryptic colouration is often associated with this behaviour, as in the spotted white-tailed deer fawn (*Odocoileus virginianus*). Some animals carry this strategy further and feign death. For example, the opossum (*Didelphis virginiana*), is harassed by a predator such as a dog, remains motionless on its back. The physiological correlates of this behaviour are not completely understood. Hog-nosed snakes (*Heterodon platyrhinos*) exhibit similar behaviour, although they may precede it by threat behaviour in which the hog-nosed snake resembles a cobra. Freezing and feigning death may work because many predators seem to respond only to moving prey. For example, wolves will not attack prey that stand motionless.

**The Development of Anti Predator Behaviour**

A baboon which is attacked by a leopard but which survives, learns something about the speed and attack methods of leopards. Future encounters with leopards will be different as a consequence of the experience. Field studies of the improvement in techniques of anti-predator behaviour are difficult, but many laboratory studies of avoidance conditioning minic certain aspects of this type of behavioural change. Suppose that a rat in a cage learns to avoid shock by pressing a set of levers in a particular sequence when a certain combination of stimuli is presented. That rat may be using abilities which would also be of use were it faced with successive exposures to a predator. The effect of experience are often easiest to study in developing animals since some aspects of previous experience can then be controlled.

In species in which parental care is prolonged, the most effective anti-predator behaviour which young animals can show is often to recognise and maintain contact with the parents and the place where they are put by the parents. The overriding importance of finding the parent leads to some apparently anomalous responses to predators. *Lorenz* (1935) describes how newly hatched precocial birds, such as goslings, show no avoidance behaviour to a variety of moving objects including, for example, man, a goose predator. Schaller and Emlen (1962) pushed objects towards the young of ten species of precocial birds and found very little avoidance reaction at 10 h of age. Avoidance in their experiments, and in those of Bateson (1964) on chicks exposed to novel moving objects in a runway, increased considerably during the first week of life but changed little after this time. The startle responses shown by young domestic chicks in their home pen, when a light-bulb on the wall was illuminated for the first time, increased in magnitude and duration with increasing age from 1 to 10 days (Broom 1969). The change with age of these studies was greatest during the first three days. As the animal gets older, it learns the characteristics of its immediate environment. If the degree of novelty of any environmental change depends upon a comparison with a 'model' of the familiar surroundigns (*Salzen* 1962), novelty will increase whilst the model is established and hence responsiveness increases with age.

A system results in the young animal showing anti-predator behaviour to any novel change in its surroundings will operate when a real predator appears. Responsiveness to potential predators by young animals does not depend solely upon degree of novelty. It is also affected by the nature of the sensory input which results from the presence of the predator. Bodily damage, a looming object, or a loud noise will elicit more extreme anti-predator behaviour than the sight of small objects, the sound of quieter noises, etc. (*Hebb* 1946). The existence of perceptual analysers in the brain which enable an individual to recognise dangerous events, for example, a looming object. Some of these analysers are known to develop in a different way according to specific visual experience by others probably depend on general environmental factors such as light and metabolic stall at key times during embryological development. The average individual, therefore, would have some predator detection mechanism whether or not there had been previous visual experience of the predator. An elaborate and specific possibility is a detector for flying bird of prey. Lorenz (1939) and *Goethe* (1940) reported that young precoeial birds did not respond to flying geese but showed a flight response when a hawk or falcon passed overhead. Experiments by *Schleidt* (1961) showed that the result could have been due to habituation to the geese but not to hawks. Further work on nai mallard ducklings by *Green, Oreen* and *Ca* (1966) and by *Mueller* and *Parker* (1080), however, indicates that hawk and goose silhouet can be distinguished and do elicit different reactions. The reactions were not clear anti-predator reactions in either case. The lining of the perceptual recognition to the command which initiates such reactions, therefore, may require experience which the experimental subjects of these experiments did not have.

The general effects of experience of anti-predator behaviour have been examined in many experiments with young animals. If the occurrence of such behaviour depends, in part, on the gradual formation of a model of familiar surroundings, with which future events can be compared, individuals reared in different surroundings will form different models which may be formed at various rates. Experiments with rhesus monkeys, chimpanzees and domestic chicks indicate that the magnitude and direction of responses to novel objects or novel environmental changes is less for individuals with more complex rearing conditions. The large literature on the effect on later behaviour in a strange pen (open field) of handling rodent pups during infaucy, leads to the same general conclusion. It seems that the major effect of handling a rodent up is to change the amount of time that the mother spends licking and sniffing that pup. Perhaps as a consequence of this, handled pups show a less marked response in novel situations. The extent and type of maternal care also has effects on the responses of young primates confronted with possible danger. For example, *Ainsworth. Bell* and *Stayton* (1971) found that one-year-old human infants, whose mothers were inadequate in their responses to them explored less and fled to mother, mother more readily in strange and hence potentially dangerous situations.

6

# Singing Behaviour

Is singing inherited or learned, nature or nurture? Some of the first attempts to answer this question were made by professor W.H. Thrope at Cambridge University. He was one of the first ethologists to use the sound spectrograph, which had been developed for analysing human speech. It was with the advent of this machine that a whole new window was opened on the study of animal sound communication. A simple way of checking for evidence of learning is to isolate hand-reared birds from any sounds that they might learn. Thrope's early work was with chaffinches. He found that birds kept in soundproof boxes were unable to sing properly. They would attempt to sing normal song, but unless brought up in the company of other normal birds, they could only attempt rather raucous noises quite unlike normal songs.

Male song birds sang abnormal songs if deprived of an opportunity to hear adult song during the first nine months of life. But, if played recorded tapes of chaffinch song, they would learn to sing accurately as early as ten days after hatching. It became clear that songs are learned. At first Thrope thought the learning process was progressing throughout the first year of a bird's life. It wasn't until later than peter Slater and his colleagues at Sussex University demonstrated that fledglings could learn whole songs, thereby narrowing down the likely learning period. Professor Peter Marler, now at Rockefeller university, New York, first worked with Thrope's chaffinches at Cambridge. He then went to the University of California at Berkeley and together with Mayukasu Konishi, now at the California Institute of Technology, Pasadena, repeated the isolation experiments, but this time with local white-crowned sparrows *Zonotrichia leucophrys* an especially common species in the San Francisco Bay area which shows very distinct local dialects.

The white-crowned sparrow has just one song type, consisting of two parts. There one or more introductory whistles followed by some brief, rapidly repeated syllables. Sometimes there is a whistle of buzz at the end. The trill portion contains the dialect are in the Bay area just by stopping for a moment and listening to the nearest white-crowned sparrow. Marler was able to confirm the chaffinch findings–white-crowned sparrows in social isolation developed abnormal songs. Although the song had the whistle sections present, the trill was absent. But there was a problem. Although the song developed by a male in social isolation is abnormal, in the sense that it sings something not heard in the field, there were still normal qualities to the song. Marler played these sounds to ornithologists and asked them to identify the species.

Invariably they would suggest that the song must be a new species of *Zonotrichia.* Marler was puzzled–just what was isolated male is still able to develop? Konishi decided to look at the effect of deafening on song production, and established that the information enabling the socially isolated birds to show a hint of normal song lies somewhere in the auditory part of the brain. A deafened male develops a song which is even more remote from normality, more primitive than anything than an intact bird will develop, even in isolation. The song of an early deafened white-crowned sparrow is a raucous buzz, more like the sound of an insect than of a bird. Marler and Konishi developed the result into a new interpretation of the song learning process–the auditory template theory of song learning. They concluded that each song bird studied has some innate capacity to develop at least some of the normal features of the song heard in the wild, and that capacity rests on a mechanism associated with the senses of hearing. When a young male begins to sing, in the normal course of events, he listens to his own voice. He has an image of what it should sound like and he proceeds to match his singing to this expectation template fashion. He gradually achieves a more perfect match with the template until he is able to sing a song with some of the normal features. In a bird hearing normal song, the template is modified, overridden by the learned information, until the song includes features of the local dialect. The male, now, has a new template which is more advanced than the innate version. He proceed through this matching process, through sub-song (an immature form of song with impure notes delivered more quierly than full song and in longer phrases) and plastic song (the pre-adult song when phrases are shortened and notes perfected), until the song crystallises into an imitation of the local dialects to which the male was exposed when he was young.

White-crowned sparrows, chaffinches and other male song birds that have been studied seem to have sensitive period for song learning, typically when they are quite young. In the white-crowned sparrow the period is somewhere between ten and 50 days, at a time when the bird has left the nest but is either dependent on or associated with its parents. The young male white-crowned sparrow tends to move away from the area of hatching somewhere around two months of age. This, then, is a case where the youngster does his song learning while still at home, although he does not start to sing himself until some time later. When he does start to sing the following spri:ıg he generates an imitation to the local dialect. The imitation. Though, is not completely faithful ton the models heard.

A young male will extract certain features from those heard in infancy and keep them intact, while allowing himself freedom to improvise or invent other portions of the song. In the relatively simple song of the white-crowned sparrow there is a strong tendency to confirm to local fashion, although each male includes individual, personalised features. Each male has set of structural features that label him 'white-crowned sparrow'. They identify him as a member of a particular species. There are some features of local dialect which tell about the area in which he was brought up. In addition, there are portions, usually in the whistle section, which are his personal marks. These sound cues are employed in different ways in the social life of the birds so that they are all know there immediate neighbours, can distinguish neighbours from more distant members of the community, can discriminate between dialects, and can identify members of there own species and other species. There is a hierarchy of song features which is present even in the relatively simple song of the white- crowned sparrow.

Sensitive periods for song learning are by no means universal. Those birds with well-defined dialects tend to have narrowly defined periods for learning; others have different learning patterns. In some, the programme of song learning remains open, sometimes for quite long periods and, In a few cases, throughout life. The red-winged blackbird continues to add to its repertoire as it matures, and the domestic form of the canary discards quite a large proportion of its song form one year to the next, adding new phrases to replace the old. In both these species the size of the song repertoire increases annually as the male gets older and more mature. Females chooses the older, fitter males, which have survived several breeding seasons, in preference to the younger, untested juveniles. In Wytham wood, Oxfordshire, John Krebs and his colleagues, notably Peter MacGregor, also directed there attention to song learning periods. They wanted to find out at what time of year the young male great tits learn their songs and from whom.

Traditionally it was thought that songs were learned from fathers. Laboratory studies with zebra finches *Poephila guttata* by Immorman and with bullfinches *Pyrrhula pyrrhula* by Nicoli, both from west Germany, showed quite explicitly that caged male birds learn songs from the father. Again, they used Chris Perrins' population of individually ringed great tits, many of whom are of known genealogy, and were able to look at the repertoires of fathers and of their offspring in the following year. The reason the Oxford researchers decided upon field study is that laboratory studies do not offer a young male a choice of different song models from which to learn. The bird sits in cage, perhaps with his parents or may be a tutor tape. In the wild a young bird sits in the nest, or files around after leaving the nest, and hears a variety of different songs. Does he know which song is his father's and then learn specifically his father's song or does he just pick up whatever he hears?

Looking at the songs of known fathers and songs, Krebs and MacGregor demonstrated that there is no tendency at all for young male great tits to learn songs form their fathers. Songs, in this species at least, are not passed from generation to generation within a family. So, from whom *does* he learn? Further investigations revealed that the young male shares the highest proportion of his songs with his neighbours in the year that he sets up his first territory, that is, usually, when he is about nine months old. If the bird is hatched in June he will become a territorial male in March the following year. This suggests that, in great tits, quite a lot of song learning occurs in the first spring of life, rather then during the first few weeks as laboratory studies of other species would imply.

Not *all* the song elements in a individual's repertoire are learned from neighbours.

In Wytham Wood some birds are 'resident' having been hatched in the wood, with known parents, and others are 'immigrants' having been raised outside the wood, which have wandered in from may be several miles away. The elements of song repertoires are derived, seemingly, from the two groups. There are the 'common variants' which many different males in the wood include in their song repertoires, and the 'rare variants', which are sung by a few individuals in the population. The 'rare variants', it turns out, tend to be sung by immigrant birds which arrived as first-year males. Since the birds have predominantly 'Wytham Wood songs' but with the occasional song type from elsewhere, Krebs and MacGregor concluded that the local 'common' song types were learned on arrival at the wood and the 'rare' song types were learned before they dispersed from their home area.

So, from the field studies on great tits, Krebs and MacGregor have identified at least two periods when the birds can learn songs: some time early in life before they disperse at the age of two or three months, and some time later at six to nine months when they first set up territory. After the first year the repertoire is not modified. Krebs and MacGregor followed individuals for several years and found that they kept exactly the same song repertoire after their first spring. Peter Slater, of Sussex University, working with captive chafinches, has results with fit with the same conclusions. He has shown that hand-raised chaffinches can learn songs, both in their first few months of life, which corresponds to the pre-dispersal learning of wild birds, and in their first spring, corresponding to post-dispersal learning.

Krebs has also used the great tit data of songs of individuals with known family histories to find out whether the songs that a female hears early in her life influence her later behaviour. Female great tits, like many female song birds, don't sing very much, but that is not to say they do not know something about the characteristics of the song. In some species of song birds a female bird injected with male hormones will begin to sing a normal male song. The song is tucked away in the brain even if it is not normally sung. Do female great tits learn songs while still very young, particularly their father's song, and then use this information later on in life, when they pair with a male, to avoid an incestuous mating with the father? Chris Perrins' analysis of reproductive success have revealed that in breeding great tits is disadvantageous. Birds mating with close relatives are less successful in breeding than individuals mating with distant relatives or individuals from outside the population. Do the females, then choose males with different songs from their fathers?

Krebs and MacGregor collected data for females where they knew the song repertoire of the female's father and the female's husband. Then they looked at the amount of sharing of songs between the two and compared the result with a chance expectation based on the rate of sharing of songs for the population as a whole. They found that females are less likely than expected by chance, to mate with males who share most of their repertoire with the female's father. In addition, they found that females are likely to turn down suitors where the song is too different from that of the father. Female great tits prefer to mate with males with song repertories slightly different from the female's father, but not very different. This result fits with the notion put forward by pat Bateson of Cambridge University that females, in choosing a males with which to mate, avoid close genetic relatives and incest, but at the same time avoid out-breeding with genetically very distant males.

Genetically very distant males may be adapted to different environmental conditions, so if there is any local population adaptation to different habitats, it would be advantageous for females to mate within their own population. The same is true for local dialect–different dialect groups might be adapted to different habitat conditions. Between learning and the fully mature song is period of great interest to researchers. This is when the learned song may change by identifiable stages, through sub-song to plastic song and finally adult song. Sub-song is rather like the babbling of a baby before it begins to talk, and as any mother knows it is a fascinating stage of development. But, it is also an almost impossible state to understand to even document because the sounds are so variable. At Rockefeller University, Peter Marler, together with Susan Peters, undertook the daunting task of analysing sub-songs, this time with song sparrows *Melospiza melodia* and swamp sparrows *M. georgiana.*

Like the white-crowned sparrow of California, swamp sparrows also learn from memory. They learn their songs when quite young; it is all over by 60 –70 days of age. They do not come into full song until, perhaps, 200 days. Subsong begins earlier, sometimes even when song learning itself is taking place at two months. It was thought traditionally that sub-song was a period of rehearsal, when songs are committed sub-song was a period of rehearsal, when songs are committed to memory as they are being heard, so that when the bird comes into full stored. Marker felt sub-song is so unstructured that it was unlikely that rehearsal was taking place. He set out to analyse it.

A group of 16 birds was set up ad recorder once a week for a year. Marler and his colleagues analysed the whole transition from sub-song, to plastic song, to adult song, and surveyed this enormous mass of material for evidence of rehearsal of recorded tapes to which the young birds had been exposed. As this was a laboratory study, Marler knew everything that the young birds had heard during infancy. He could look through the record for the very first appearance of anything that was remotely like the structure of the models that had been learned when the birds were young. Rehearsal, it turned out, did not begin until the birds were between 230 and 250 days of age. The male swamp sparrow had learned the tape recorded songs when they were young committed them to memory by ear, without rehearsal, and stored them for up to 300 days before they began to produce anything remotely resembling the original model–a quite extraordinary feat. One of the targets of Marler's current research is to look at the form in which the birds commit the songs to memory.

Birds, Peter Marler suggests, have an ability to memorise a model when young, and break it up into sections, or syllables, which can be rearranged to create new songs. Birds, then, have the capacity to retain phonetic units that conform to the local fashion and yet can be recombined in new ways, allowing each male to place his personal mark on the song. They do this so freely that it is tempting to think that, when the song is committed to memory, it may well be done in a segmented form. By breaking the songs down into sub-until a bird could easily combine them later when it starts to sing in earnest. The units, interestingly enough, are not the smallest units of a song. Birds, it seems have a remarkable capacity to make use of something which is very much like the syntax of language. This is not the 'lexical syntax' of a human sentence; the components of bird song do not have independent meanings in the same way as human words. The song is a message as a whole. There is however, another level of syntax called 'phonological syntax'. Combinations of consonants and vowels, or 'phonemes' from which words are constructed in human speech would be analogous to the syllables of bird song. For humans every language has a limited set of phonemes and an infinite number of ways in which they can be combined, giving an endless capacity to generate new words.

This simple song of a sparrow has a dry repetitive trill, with a syllables repeated perhaps 20 times. Each of these syllables may be made up of six or seven distinct notes. The songs may be made up of six or seven distinct notes. The songs are broken down to the level of syllables and not notes. Marler feels this is. Perhaps, a natural unit for the production of sound–the way in which some linguistic view the phoneme. In addition to the imitation. Marler's swamp sparrows also invented song material. As he had such a complete record of everything the birds had heard, Marler was able to direct these subtle variations. When the birds sang their full song the researches would to through the spectrogram item by item, first looking for events attributable to learning in infancy. They discovered that some of the models. The components in adulthood did not match any of the models. The components were seen as copies at an intermediate stage of development but then they were changed. The bird had engaged in a kind of vocal play, imposing small transformations, and so getting further and further away form the original model. By the time bird produces adult song the end product cannot be related easily to the starting point.

The red-winged blackbird is an obsessive 'inventor'. It is very difficult to identify imitation until the intermediate developmental stages are examined. Some birds generate seemingly completely original material that is not attributable, as far as can be seen, to the features of the original model. In the original Cambridge, chaffinch days the process of imitations was emphasized in song learning. The picture we now is a lot more complicated. Along with imitation there is improvisation and invention going on. Birds are constantly generating new patterns of sound in nature.

One of the fascinating things Marler and his colleagues have discovered when analysing sub-song and plastic song in the swamp sparrow is that each bird generates far more song material in the course of development than is needed for normal adult song production. A male swamp sparrow has, typically, three song types.

Marler analysed the plastic song a week before it had been using six different syllables. A couple of weeks before it had been using 12 (in one case as many as 19) As it approaches the moment of song crystallisation, the song components get winnowed down very until the mature repertoire is left.

In a male producing 12 syllable types, Marler found that five or six were obvious imitations, one or two were improvised variations of an imitation, and four or five seemed to have been invented. Even though he had examined carefully the development record, he could not find ancestors in the learned song, nor were the invented syllables like the sounds produced by a socially isolated bird which has had no opportunity to learn. The syllables in 'isolate; song are simple, whereas the invented syllables are complex, although not as complex as the imitations. Marler concluded that the bird has a capacity for a truly creative process. He feels though, that the creativity is not realised unless they are pushed over a certain threshold, as though their environment must in some way be enriched to a degree that motivates them to indulge in this creative activity. This is the target for future research-what kinds of conditions provoke creative invention?

The problem confronting. Marler is not new. Much psychological researches indicates that enriched environments are important to young organisms, not just to provide models for imitation, but to stimulate them to initiate new activities Marler is tempted to strike an analogy with play. The vocal gyrations that birds go through during late sub-song and early plastic song are, in effect, vocal play, and play behaviour is one of the patterns of behaviour known to be influenced by the richness of the environment. This is particularly true in children. As the invented syllables are a very recent. So far they have discovered creativity in the song of a bird with one of the simplest of songs, the swamp sparrow. It is exciting to contemplate the degree of inventiveness to be discovered in the songs of mocking-birds or blackbirds. This is a promising area for future research.

If all song birds are learning songs at roughly the same time, why don't some of them learn the wrong ones? In the early days of research, the Cambridge chaffinches were presented with the song of the tree pipit *Anthus trivialis* and learned it. This was pursued further, by Peter Marler, with the white-crowned sparrows in California. They were given a natural choice of songs learn. In the coastal in which white-crowned sparrows live, another very common species is the song sparrow. The two species live closely together. As far as field research could show, white-crowned sparrows never song sparrow. Birds were brought into the laboratory.

White-crowns were given a choice of recordings of white-crowned and song sparrows and unerringly imitated white-crown.. If they were given song sparrow alone, the birds would reject it and revert to their innate song. Marler was intrigued. Here was one of the most elaborate learning process known in animals and yet there appeared to be innate instructions or guidelines as to what the bird should be learning, an interest interplay between nature and nurture. A bird cannot develop normal song without the opportunity to learn and yet it inherits some capacity to identify what it ought to be learning. Marler and Susan Peters set out to see whether the swamp sparrows and song sparrows of New York had innate guidelines for song learning. Again, these two species live closely together in the same habitat, without earshot of each other. Instead of playing them tape recordings of natural songs, marler began to synthesise songs using a computer.

Syllables from natural songs of the two species were edited out and recombined in different patterns, some swamps sparrow-like, others song sparrow-like. For every patterns devised two versions were created–one composed of mainly swamp sparrow song and the other of song sparrow. The research team gathered its battery of training songs, brought birds into the laboratory, and trained them between 20 and 60 days of age, the period when their readiness to learn is at its maximum. Then they waited to see what they would copy from the training songs. The results were interesting. Take the swamp sparrows, for instance. Marler found that they wee highly selective in what they would learn, and the selectively was based, not on the pattern ,but on the syllable irrespective would learn swamp sparrow syllable irrespective of the pattern in which it was presented.

This was the first robust demonstration of selective learning as an all-or-nothing phenomenon. Marler is now attempting to pin down the acoustic features on which a swamp sparrow makes its innate choice of what it should learn and what it should not learn. He has identified already the features that pitch between the two species but manipulating and swapping the frequencies did not make any difference.

Might there be something about the vocal tract would make it physically difficult, if not impossible, for a swamp sparrow to sing the same syllables as a song sparrow? Marler devised a way of testing for this. It is possible to get a song sparrow to learn a song which has song sparrow patterning, but which is made up of swamp sparrow syllables. Marler took a swamp sparrow syllables and taught it to a song sparrow. He could then see if the syllable had become unacceptable, having gone through the vocal tract of a song sparrow. The swamp sparrow learned the syllables.

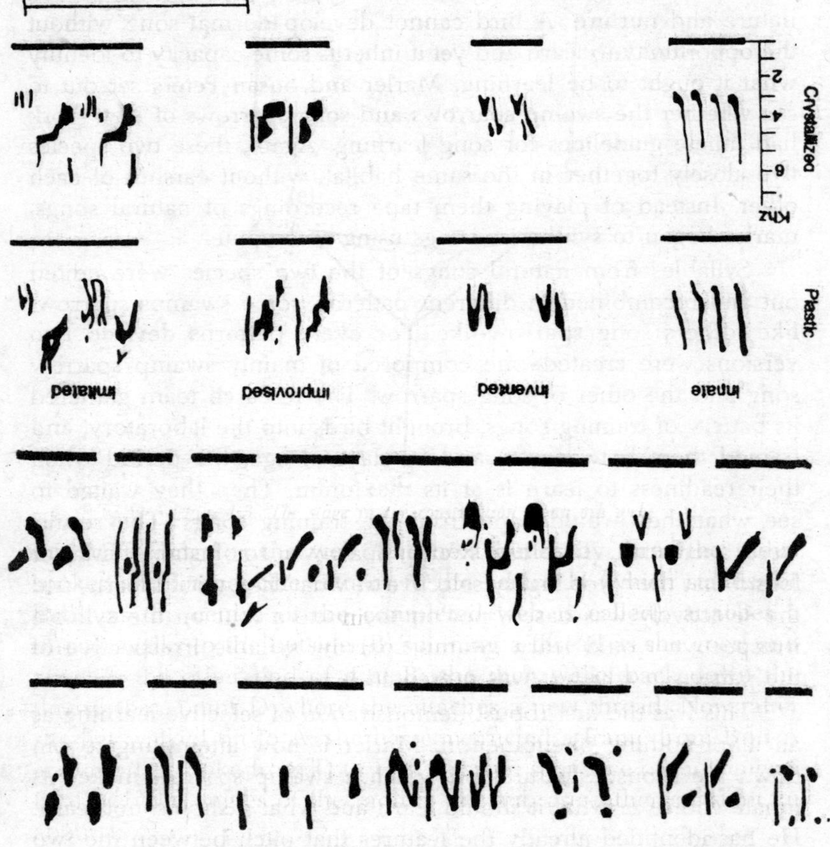


*Fig. 6.1. Pairs of song syllables used for training swamp sparrows and pairs of syllables developed later in plastic and crystallised song, classified as inventions, improvisations and imitations.*

To cut a long story short, Marler eventually concluded that an innate perceptual process must be at the heart of selective learning. Birds, from the age of a few weeks, have an innate perceptual filter in the brain which accepts or discards incoming auditory information. Other researches are not so convinced and suggest that the birds may *remember* everything but *reproduce* only the appropriate songs. More recently, Luis Baptista, experiments with white-crowned sparrows in California, has cast doubts on the universality of any innate mechanism. He succeeded in teaching the birds all sorts of alien songs which, if a perceptual filter was working in their was to put the tutor and pupil in visual contract. Birds, it seems, are fussy about what they learn from recorded tapes and are more prepared to learn from individuals with which they can interact.

Marler, however, still believes that, given the multitude of environmental stimuli affecting young organisms, they must be provided with some innate instructions about how to sift the barrage of incoming information if their nervous systems are not to be reduced to chaos. To which stimuli should the creature attend and which ignore? At which point should it be ready to accept less favourable stimuli if the optimum circumstance do not arise? There must be organism must go through a certain process of behavioural development to provide the skeleton on which later stages of development must depend.

## Song Dialects

The consequence of birds having learned their songs is that individual of the same species, living in different areas, will sing songs which are not identical. Just as local variations in human speech occur from place to place within a country sharing the same languages, so too do some birds have small pockets of song variation–local dialects. Ornithologists in the Scan Francisco Bay area, for example, can place a white-crowned sparrow to within a couple of kilometers of its home range simply by hearing its song. When at Berkeley, Peter Marler, together with M. Tamura, analysed sonagraphically the songs of white-crowned sparrows and discovered considerable variation in the sample song.

Basically, these birds have an alerting whistle, followed by a campus sing a simple series of three whistles and end in a Francisco, have an opening whistle, a trill, and finish with a buzz. Individuals from Sunset Beach to the South, have a double whistle followed

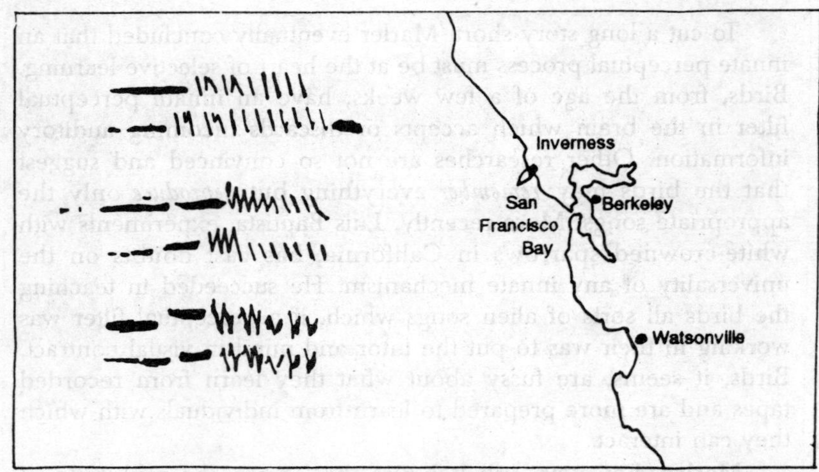


*Fig. 6.2. Songs of male white-crowned sparrows from three distinct dialect areas around San Francisco Bay.*

by the buzz and terminating in a sparrow songs. In Europe, the same distinct dialect areas have been noted for redwings *Turdus iliacus* and corn buntings *Emberiza calandra.* Marler and Tamura went further, and tested white-crowned sparrows for evidence that dialects were, indeed, a result of cultural transmission during song learning.

Hatchlings that brought into the laboratory and then reared in social isolation. The hatchlings deprived of adult example sang songs quite unlike their home dialect, whereas the birds that had heard the songs of their adult neighbours eventually sang good copies of the local dialect. Tutoring studies later showed that sparrows would learn, during the critical learning period, the home dialect or even the dialects from other areas. This was evidence that dialects are learned; but what of their function? Could it be that dialects are simple accidental occurrences as a result of learning and have no biological function? Marler and Tamura have speculated that female sparrows might use the song dialect for selecting a mate. The dialect might represent a local population indulging in some degree of inbreeding, and therefore, in a sense, could be considered as some form of incipient speciation, with the advantage that, as it is a learned behaviour, it could be reversed during the course of a generation. Unfortunately, work by Luis Baptista has confused speculations of this sort in that female white-crowned sparrows given testosterone to encourage them to sing, surprisingly sang with a different from the one in which they were brought up!

Myron Baker, of Colorado state University at Fort Collins, took another line. He wanted to see of any genetic variations coincide with a boundary between two dialects. Using biochemical analyses to detect polymorphisms in the blood, Baker examined the genetic make-up of birds across a dialect boundary. He discovered genetic discontinuities which, he argued, are consequences of males and females settling in an area where their home dialect is sung. He pursued this further by banding birds which had hatched along the transect and followed their patterns of dispersal. He found that birds in the heart of a dialect area disperse more or less radially to establish territories and breed, whereas those at the boundaries dispersed asymmetrically, veering away from the boundary into focus of the home dialect. Baker had revealed that the dialects of his study group of a sub-species of white-crowned sparrow from a kind of genetic mosaic. They are members of the same species yet there is enough differentiation for each dialect area to constitute a deme or selective unit. This would also mean that birds living close to each other tend to be kin, a situation which favours some of the elaborations of social behaviour, such as reciprocal altruism , emerging.

In the light of these kinds of observations it has been suggested that dialects do have a function. Fernando Nottebohm, now at Rockefeller University, New York, proposed a function for dialect when studying the rufous collared sparrow *Zonotrichia capensis* on the pampas of Argentina. He suggests that the female of a species such as the rufous collared sparrow which has a wide distribution in a variety of climates or habitats from cold tundra to hot desert, would benefit by mating with her own dialect group. A bird adapted to a desert would best pair with another desert bird of the same species. Mating with a bird from a polar area, for instance, would dilute the gene pool, possibly introducing inappropriate characteristics, and might make the offspring less likely to survive bears the rigours of the desert. The rufous collard sparrow bears this out. Across large tracts of the Pampas, with uniform plains vegetations, the songs of the sparrow are very similar. The birds in the hills, where the climate and vegetation change, have quite different dialects.

But, although the work of marler, baker and Nottebohm has suggested a function for dialects in white-crowned sparrows and rufous collard sparrows, Peter Slater, of the University of Sussex, is reluctant to extend these findings to song variations in many other species. Slater feels that the tremendous amount of emphasis put on dialects and geographical variations in bird song has been overdone. His interest is in chaffinches, and he feels that variations in their songs serve no function whatsoever. In the case of the chaffinch, dialect-like differences really are an accidental by-product of song learning. A young bird, within the first year of life, hears other chaffinches singing. He listens and learns, memorising the exact details heard.

The following spring, when his testosterone levels begin to rise, he starts to sing himself, matching his own output to his memory of what he has previously heard. Slater has demonstrated that hand-reared chaffinches, trained in the first two months of their life with song types that they did not hear subsequently, produced near perfect copies six or eight months later. The problem about learning as a strategy of song development, suggests slater, is that animals make mistakes, and mistakes in learned patterns are much more common than mutations in genetic transmission. As a result song is almost certain to change slowly. Some individuals copy accurately, other not so accurately. Some song types may copied less than others and become extinct; others are miscopied, so creating new songs. If populations are distinct from each other, then birds will learn primarily within their own population, and as a result the song and their mistakes will be different from those of a population some distance away.

In the wind-swept Orkney Islands, off the north of the Scottish mainland, Slater has been studying a group of chaffinches and looking for local song variations. There are few trees on the Orkney, consequently few chaffinches, so Slater has been able to record a very high proportion of all the one song type for the whole of Orkney and several song types for each major wood, but there were also individual variations. Each chaffinch has between one and five song types. There are song types shared by many birds and other song types singular to a particular bird. In one wood, for example, with 15 male chaffinches, Slater identified ten individuals singing song-type B–the phrases within the song were almost identical with few variations. But another bird might have a type very similar to B, but not quite the same. Slater suggests it had copied song type B but had made a mistake.

From the 40 male birds in the study area, Slate was able to recognise 17 different song types. If every bird had its own songs the number would have been nearer a hundred, but this was not the case. There was obviously a degree of overlap of the song types. The grouping, though, is not strictly into dialects. If it were one strand of trees would have a particular collection of song types, while the next wood would have a quite distinct set. This is not what happens. Slater found that songs tend to have foci. When the song is common about half of the population are singing it. As you get further from the song focus fewer birds are heard singing that particular song and another song type takes over. Travel a little further and you reach the focus of the new song. In chaffinches there are no distinct dialect boundaries. Rather, there is a patchwork of overlapping song types where some song foci interact with others, and where some song types are common to all areas. The song simply carries geographically from one area to another. Peter Slater feels that the word 'dialect' should be reserved for those species where there are areas in which a particular song type is sung, marked by order lines beyond which the song is not sung and another song is sung instead.

## Vocal Copying

One way in which a song bird can acquire an elaborate song is to borrow elements from the songs of others–vocal copying. After all, song birds mostly learn their species song originally from neighbours. This kind of copying from males of the same species is the first of four categories of vocal copying in song birds described by David Dobkin, of the University of California at Berkeley. He calls this kind of infraspecific copying 'vocal imitation'. It is the mechanism of dialect systems. Dobkin thought that the literature was littered with a host of interchangeable names that described any form of vocal copying–vocal mimicry, vocal imitation or vocal convergence–and was determined to clear things up in order that scientist would be talking about the same piece of behaviour. Dobkin came to this task when he became interested in the problem of two relatively closely related species living in the same area. If they are breeding within earshot of one another, using songs to keep rivals of their own species at bay, and to attract females of the right species, what are the implications for the kinds of signals being transmitted?

The birds are potentially close competitors, particularly if they are morphologically similar. They might be feeding at the same time, on the same kinds of foods and have similar nesting requirements. Species A needs not away species B males. This also raise problems of unambiguous recognition of signals on the part of each species. If species A vocalisations are similar to those of species B, then species A males are likely to attract females of species B, resulting in no offspring, or at the last, infertile hybrids. This is not advantageous, clearly, for the individual, so there should be some selection for species specificity. But at the same time, if a bird is in a excluding potential competitor of another species, then there is opposite category of Vocal copying Dobkin has called 'vocal convergence', and it occurs when ecological competitors or close relatives have similar songs or calls. Dobkin termed his third category 'vocal appropriation'. This happens when an individual copies sound made by members of another unrelated species.

It is often accidental, as when an individual incorporates into its song elements of another species' song simply because it lives in the same acoustic environment. Vocal appropriations may occur as a result of an impoverished acoustic environment. If a species is nesting in an area at a very low density, and there are there are not many role models around from which young birds can learn, then a youngster may pick up song elements of a close neighbour of a different species. In the literature vocal appropriation has often been described as 'Vocal mimicry' but Dobkin is fast to point out that mimicry involves deception and deception is probably not intended in this form of copying; it is simply a way of enriching the song repertoire.

Vocal mimicry is, in fact, Dobkin's fourth category of vocal copying, but he questions its existence in animal vocal communication. There is, he believes, only one documents study in which vocal mimicry takes place. This is the case of violaceous euphonia *Euphonia violacea,* a member of the tanager family. If a predator approaches the nest, the resident pair give the alarm and mobbing calls of other species which are attracted to the nest and mob the predator. Some birds copy bizarre sounds. Blackbirds have been heard to copy 'whistling' telephones and pedestrian-crossing bleepers. Why they should do this is a mystery. On an old Ludwig Koch recording, made in 1910, a celebrated Prussian blackbird even imitated the distinctive sound of the Kaiser's motor klaxon. The superb lyrebird *Menura novae-hollandiae* of Australia regularly incorporates farm machinery noises into its courtship display.

An even more puzzling form of elaboration to a bird call occurs when captive hill mynahs Gracula religiosa accurately copy human speech. In the wild copying is basically confined to species calls or calls of neighbours, although there are many exceptions. Mynahs often incorporate the staccato sounds of local tree frogs into their complex calls. In some species, for example tropical boubou shrikes *Laniarius ferrugineus,* an elaborate song is built up from the contributions of two birds. Instead of the male singing alone, the female joins in too. The coordinations and integration of the two songs is so well achieved that is often sounds like the song of just one bird. How the co-ordination is achieved we do not know.

Why should birds want to duet in this fashion? A clue may lie in the habitat of most duetting birds, for they live almost exclusively in tropical forests of other places with dense vegetation. They also tend to be monogamous–pairing for life–and to retain territories throughout the year, for many years. Duetting, perhaps, maintains and reinforces the pair bond. In dense vegetation with birds constantly out of sight of each other, a sound reinforcer would be important. Duetting birds are often seen singing at other duetting birds across a territorial boundary. Duetting, then, like solo song, may have the dual function of establishing and reinforcing the relationship between a pair and of territorial proclamation.

## Song Production and Control

In order to make their beautiful songs, birds employ some clever anatomical tricks. Basically, air from the lungs is forced over vibrating membranes–the syrinx–much as in our own larynx, but there the similarity ends. The birds syrinx is a little more complicated. Unlike the human larynx, which is at the top of the trachea or windpipe, the bird syrinx is at the bottom, where the two bronchi meet coming from the lungs. The syrinx sits astride the junction.

In many birds there are separate vibrating membranes of chords on the two arms of the syrinx, and as a consequences they have the remarkable ability to produce two sounds at once. A bird can sing two quite distinct tunes at the same time. Another difference from human speech production is that much of the structure of

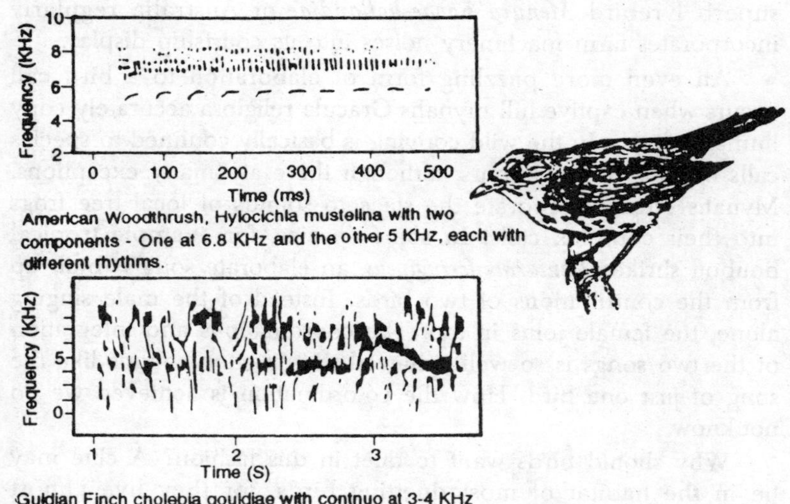


*Fig. 6.3. Complete songs of birds showing contributions from left and right half of syrinx.*

speech results not from variations in the voicebox bit from movements of the tongue, mouth and nose, which change the resonant properties of the various cavities. Bird song is not produced that way. Most birds are not dependent on the resonance of the vocal tract. Instead, the entire system is driven actively by the membranes of the syrinx. American thrushes are fine examples of birds that can sing two songs simultaneously, bit with a harmonious relationship between each rendition, producing some quite extraordinary tonal qualities which, some researches feel, are not to be matched anywhere else in bird song. A mature song bird, then, has sets of contributions to its song–one coming from the left side of the syrinx and the other form the right–and they are precisely co-ordinated. One of the problems a young bird must overcome is how to achieve that coordination. Fernando Nottebohm, of Rockefeller University, New York, studied the nestlings calls or the fledgling calls produced by a nestling chaffinch, and could distinguished between the contributions from the right and left side. He found that, in the young bird, the relationship between the two sides of the syrinx is not controlled.

The coordination varies, giving the song a harsh quality. There is evidence that the fledgling call of the chaffinch provides practice in syrinx coordination, which is a necessary prelude to the process of singing and of learning. Nottebohm followed the work through with domestic canaries and showed that there is a parallel between bird song production and human speech production in that one side of the syrinx tends to make a dominant contributions. In the case of the canary, it is the left side. And the dominance is not only evident in the syrinx. Dominance of the left side can be traced up into the brain–an analogy with the phenomenon of hemispheric dominance in the control of human speech. This is the only animal analogue for this phenomenon studied so far and it has been the thrust of Fernando Nottebohm's research for the past ten years.

Birds, we have seen, learn to sing by reference to auditory information. There is a set of built-in rules that specify the kinds of auditory in formation to be accepted and a repertoire is developed by reference to these auditory expectations. Normally the song manifests itself in imitations of other adult models. How, then, is the brain handling this? In Principle, there should be an efferent motor system and an auditory feedback, which at some point must integrate, so that when the song is matched to the auditory template, and learning completed, the pattern may be held and stored in long-term memory. Nottebohm wanted to find out which part of the brain carried out these tasks. He started by looking at the motor part of the auditory loop as that was more accessible. Attention was focused on the syrinx.

Nottebohm and his colleagues Christiana Leonard and Tegner Stoked, discovered three regions of nerve cells in the brain which seem to the devoted, almost exclusively, to the control of syringeal muscles, and which he assumes also control aspects of song learning related to the performance of the syrinx. These song nuclei are simply clusters of cells which are quite distinct, with discrete boundaries and which can be easily identified with specific staining techniques. Two are located in the forebrain, the telencephalon, and one in the hypoglossus. The hyperstrianum ventral parts caudale known as the HVc at the top of the (RA), which in turn tells the neurones that enervate the muscles (hypoglossal motor neurones) to organise and activate the muscles of the syrinx.

Nottebohm carried out some simple surgery on the brains of canaries. If he lesioned the top nucleus, HVc, the birds would continue to sing, but they did so in a virtually silent fashion. The dynamics of song were expressed but not the sound itself. The throat quivered, the bill held slightly open, the body feathers were sleeked and the bird turned around and directed its behaviour to another individual in another cage. All that was heard was a faint clicking sound, which corresponded to the temporal pattern of the song, but there was no frequency modulation, none of the sophisticated patterning characteristic of canary song,. It was, Nottebohm recalls, as if the motivation for singing was there, much of the motor control for song was present, but the syrinx did not function. Nottebohm had demonstrated that the three-nuclei auditory pathway is primarily concerned with song and call production.

In canaries, it is mainly the left half of the syrinx that is involved in singing. The right half does very little; indeed in some species it does not play any role in sound production. The nerves running from the brain to the syrinx and back are ipsilateral, that is, the song nuclei in the left side of the brain control the left side of the syrinx. This is unlike other motor pathways, which are crossed, so that one brain hemisphere controls the behaviour of the other half of the body. In humans, for example, the left hemisphere controls what the right hand is doing. In canaries the left hemisphere is dominant for song production.

In song nuclei on the right side of the canary brain are lesioned singing is barely affected: some syllables tend to be unstable and a few disappear, but the phrase structure and repetition of syllables typical of canary song persist. There was, however a surprise in store for Nottebohm and his colleagues. Although there is marked dominance of the auditory system by the left side, if a canary is left for a few months, gradually the right side will take over. So, even though the bird has two parallel pathways for songs control, and even though the left side is dominant, it has a right side which is underused, but can do the job just as well. Left hemisphere dominance seems to be the case also in chaffinches, white-crowned sparrows, Java sparrows *Padda oryzivora* and white-throated sparrows *Zonotrichia albicollis*. But not all birds are left-handers–zebra finches appear to be right-handers. Within a species there is a strong tendency to consistency.

If brain nuclei control a bird's singing abilities, an activity taking place mainly in spring and early summer, it might be expected that the nuclei change with the seasons Nottebohm looked to see if canary song nuclei alter during the course of the year. He found that in the spring, when canaries sing a lot, the nuclei are very large. In the autumn, five months later, at the end of the moult, the nuclei are half as large. The observations are relatively new and go on further, but Nottebohm speculates that the reduction in size is linked to a reduction in the amount of time devoted to the control of to a reduction in the amount of time devoted to the control of singing behaviour. He suggests that the network is partially dismantled.

Micro anatomy by Timothy De Voke has shown that the dendrites, the fine processes on brain cells, are absorbed. This is in line with observations that a canary learns a completely new song repertoire each year Maybe, proposes Nottebohm. The bird is rejuvenating its brain pathways for song control, so that when it starts a new episode of growth, the network is ready for song learning once more. The researchers are now looking for the exact point during the five months between spring and autumn when the changes take place. The hunch is that it could be closely associated with the moult, when all kinds of physiological changes are taking place.

Nottebohm is tempted to think to the moult as being a time when extra network space is thrown away to make room for the new wave for learning. It might also be expected that song nuclei would be of different sizes in males and females. Arthur Arnold, working with Nottebohm, found that the song nuclei of canaries and zebra finches, particularly those in the forebrain were several times larger in males than in females. At the time, the orthodox view was that the brain of males and females in any vertebrate species were virtually identical. Only small differences, requiring microdissection and electron microscopy to detect, were thought to occur. In the canary the sexual dimorphism is so marked that a slice of brain tissue, held up to the light, will show the difference clearly if the nuclei are stained a dark colour. This seemed to make sense to Arnold and Nottebohm. The male canary spends a good deal of time learning to sing and then sings in the spring. The female does not sing at all. It looked like good brain economics that if the female does not indulge in a particular piece of behaviour then se should not devote much brain space to its control.

The researchers went on step further. Male canaries have great variability in their songs. Some have song repertories three times larger than others. Arnold and Nottebohm looked to see if repertoire size was correlated with nuclei size. They found that birds learning larger song repertories have larger forebrainsong nuclei in the brain. Nottebohm calls this the 'library principle'. He suggests that learning is like the storing of books in a library. If you want to store a lot of books you need a lot of shelf space. You may, on the other hand, have a lot of shelf space and few books. Size, he feels, acts in a permissive manner. If a lot of skill for a particular problem is developed then it helps to have a lot of brain space devoted to the task. If the brain space at the outsets is small, then the chances of excelling are small. Interestingly, female canaries can be induced to sing with hormone treatment. Three weeks after injection of the male hormone, testosterone females will begin to come in to song, but the complexity of the song is that which would be expected from the small size of their song nuclei. Instead of 30-35 different syllables, they only sing five to ten different syllables types.

One other interesting observations is that after testosterone has been given, the dendrites of the brain cells in the song nuclei proliferate, presumably adding to the network space available in the brain to control the new behaviour. Nottebohm postulates that, in spring the male song nuclei also increase the number of dendrites with the secretion of testosterone. He also speculates that there may be an increase in the number of actual brain cells, a revolutionary proposal. An increase in the number of cells has been found but further work must be carried out to as certain whether they are neurones or other types of cells.

# 7

# Bird Calls

In an attempt to identify the relative importance of sound and vision in sea bird recognition, Professor W.H. Thorpe of Cambridge University a pioneer in the study of animal sounds visited the Bass Rock in the firth of forth. There he was confronted with an enormous colony of noisy gannets *Sula bassana*. Even in the gannet's simple squawky call, Thorpe found that there was enough information for the individual to be recongnised by others in the colony. While recording gannets on their craggy cliff-top nests he noticed that, when one of a pair returned with its catch, it would fly to the bottom of the cliff, hang in the up-draft and move up the cliff faces as if going up in a lift, calling all the way. The gannet on the nest paid no attention to the rising current of birds until it heard the returning mate. It became exited, and began calling back. Thorpe concluded that each bird could recognise calls of its own mate as distinct from those of others. He found that only the first part of the call, the first tenth of second needed.

A similar study was carried out with sandwich terns *Sterna sandvicenis*; The parent terns locate the nest site with aid of their visual memory. They give a 'fish call' when approaching the area, but only their own offspring respond. The young quite obviously recognised, Thorpe discovered, three distinct sections in the call of the parents. One of the first scientists to show that birds were able to recognise each other by sound alone was B. Tschanz. In 1968, he published a paper which showed that individual guillemots *Uria aalge* produced physically distinct calls to which their young or mates reacted significantly more than to the calls of others. Tschanz

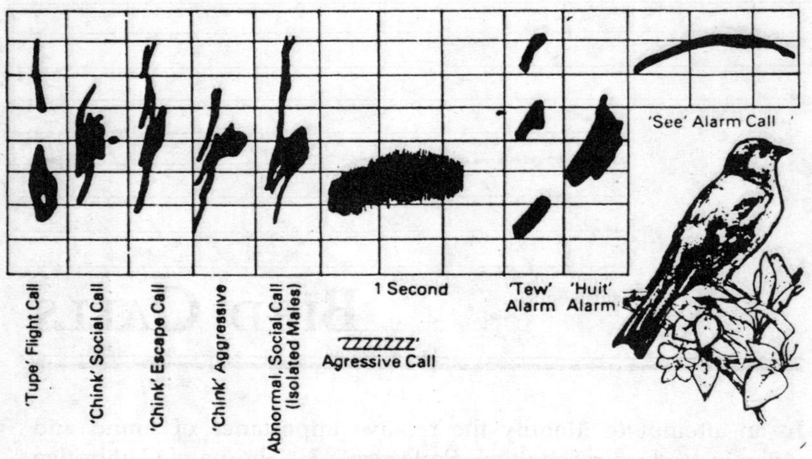


*Fig. 7.1. Chaffinch call sonagrams.*

carried out playback experiments. Young birds in the nest were exposed to parents calls and control calls alternately and simultaneously, using two speakers.

Young guillemots paid no attention to the control calls and responded only to the calls of their parents. Indeed, they would peck, at and beg from the loudspeaker playing parental calls. Tschanz also went back a stage in the birds' development and exposed eggs in incubators to particulars adult calls. After hatching they were the adult calls and other control calls. They responded to the calls to which they had been exposed while in the egg. An auditory signal can be even by the embryo before hatching. Individuals recognitions is only one function of bird calls. Social calls may keep a flock together. Alarm calls warn others of danger. Feeding calls are exchanged between mates. There may be roosting calls, nest site calls, and so on.

Calls can be associated with a particular events in the birds life. Aggressive situation might give rise to a threat call. Youngsters give begging calls for the food. Thrope worked out the vocabulary of calls for variety of species of birds, and found that the maximum number of calls a bird might have is around 15. The passerine birds came top of the list. Bird calls tend to be simple, almost monosyllabic, and relatively easy to understand. A blackbird gives a 'pinking' call in response to the presence of the neighbourhood cat. The meaning is clear.

| *Name of call* | *Call* | *Remarks* |
|---|---|---|
| 1. Fight call | 'Tsup' or'Tupe' | A short penetrating call. Low pitch with single higher harmonic. Associated with flight or preparing for flight. |
| 2. Social call | 'Chink'or 'Spink' | Clear ringing call in two parts. Helps separated birds to meet again. |
| 3. Escape call | 'Cheenk' | More shrill than 2. used as escape call and during courtship by newly paired males. |
| 4. Aggressive call | 'Zzzzzz'or'Zh-zh-zh' | Low buzz uttered during attack and fighting in a few captive males. |
| 5. Alarm call | 'Tew' | Most frequent of three alarms. Common in young birds and more rarely adults of both sexes. |
| 6. Alarm call | 'Seee' | Extreme alarm in breeding male. Pure tone rising and failing and difficult for predator to locate precisely |
| 7. Alarm call | 'Huit'or'Whit' | Commonly at rate of 30 per minute. Male chaffin ches in spring. Moderate danger. |
| 8. Injury call | 'Tsece' | Squeak given by birds hurt fighting. |
| 9. Courtship call | 'Kseep', 'Tsit', 'Chwit', 'Tzit' | Short and high pitched in bursts of three simultaneous descending notes. Male gives it courting female at pair formation in early part of season. |
| 10. Courtship call | 'Tchirp'or'Chirri' | Coarse chrip. Replaces 9 during courtship later in season. |
| 11. Courtship call | 'Seep' | The only call of female during breeding season. Short and high-pitched but composed of only two notes. |
| 12. Begging call | 'Cheep' | Soft note of nestlings. |

| | | |
|---|---|---|
| 13. Begging call | 'Chirrup' | Load and penetrating call of fledgelings. |
| 14. Intermediates | 'Huit' 'Seek' 'huit', 'Chink' | Occasional intermediates between alarm and social calls of mature birds. |
| 15. Sub-song | 'Chrip'and variants or low-pitched rattles grouped together in various ways | Chrips and warbles of birds in first summer. |
| 16. Song | 'Tchip-tchip -tchip-cherry-erry -erry-Tchip- Tcheweeoo | Tribes and a terminal flourish lasting sone two to three seconds. January or February February to June, and September and October. |

Most calls are thought to be inherited rather than learned, and to be passed on unaltered from generation. Recent research, however, has revealed that things are not always as simple as this. Paul Mundinger, at Queen's College, New York, studied the North American goldfinch *Carduelis tristis.* One of its calls, a flight call, is used by members of a pair to maintain contact during the breeding season. The female goldfinch has the habit of incubating the eggs for long periods of time. The male feeds the female. As soon as the male returns to the nest area, the female begins to beg. Mundinger, hidden near the goldfinch nest, noticed that the female began the begging behaviour long before she could see the male. He concluded the female had heard the flight the call that the male produces during its 'bounding' flight. Mundinger recorded the flight call and analysis revealed it to be more complicated than many other birds calls. In addition, individuals had subtly different variations.

The next spring, Mundinger was able to show that, at the beginning of the breeding season, a pair of goldfinches will modify the structures of their flight calls to a common pattern. One bird, the male, in fact, imitates one of the calls of the other. They maintain this pattern throughout the breeding season, using the calls as kind of naming system. He was also able to demonstrate that this capacity to learn new flight calls is maintained throughout life. Mundinger had male goldfinches as old as nine years still learning new flight calls when each new breeding season came around. The advantage to the male dies or moves away. Further research calls in the repertoire are learned. Finches reared in acoustic isolation produce calls which seem normal to the human ear. The modifications through learning are quite minor and involve detail in call structure. Nonetheless, the modifications are sufficient for individual finches to discriminate one another's calls.

In Norway, Paul Mundinger and Peter Marler discovered similarly, that mated pairs of twites *Acanthis flavirostris* have calls in common, which are different from the calls of other pairs, and which appear to be learned. They use the calls to make contact when in flocks. At the University of Sussex, Peter Slater noticed a similar learned component in certain chaffinch *Fringilla coelebs* calls. In some of his hand reared chaffinch the 'chink' call is quite different from the same call given by wild birds. Having not heard the Chink call of others, the hand reared bird is at a disadvantages and consequently produces ill-rehearsed and variable sounds. Luis Baptista found dialects in chaffinch rain calls; birds in separate areas gave different calls. Peter Slater had a laboratory chaffinch duetting with a sparrow outside the building. The chaffinch would give the sparrow 'cheep' when the sparrow did so. Clearly, birds are able to copy calls, even the calls of other species , and learning through copying is obviously necessary for the proper development of calls within a repertoire.

Calls between different species can sometimes convey information that would benefit individuals of other species. Alarm calls are obviously useful to all listeners. But the classic example of a bird using a particular call to attract, not a bird, but other animals, is the African honey-guide *Indicator indicator.* With a special call, the honey-guide leads animals such as honey badges (ratels), or even local tribesmen, to the nest of wild bees. The honey-guide is not able to break open the honey bee nest and so it enlists the help of a larger animal. The ratel and the honey guide were thought to be the original members of this symbiotic relationship. Man took advantage later. There is some evidence that in certain more urbanised areas, honeyguides have given up trying to entice men to help them, and concentrate solely on the badger population fir their honey supplies. Honey, apparently, is more easily obtained at the local supermarket.

**Echolocation in Birds**

Some birds, like bats and dolphins, indulge in auto-communication, communication with self. They bounce sounds off obstacles or targets in the environments in order to orientate and navigate. The oil bird *Steatornis caripensis* of South and Central America and the west Indies is a kind of nightjar that roosts in caves. Its nest site is located on the cave wall. Each evening birds fly in and out of the cave walls enable them to find their way in the dark. Each click consists of a burst of pluses, the separation of which we cannot appreciate until the call is analysed spectrographically. The clicks are not ultrasonic and can be heard by the human ear. The created by a disturbed colony of oil birds in a cave is deafending.

In experiments in darkened flight cages, oil birds have been shown to use their clicks, as do bats, to avoid obstacles in their flight path; if their ears are plugged they crash into things. When executing a complained manoeuvre, such as landing on the nest site ledge, bird's click rate increases, in the same fashion as the calls of the vespertilionid bats; the more clicks, the more information gained from the returning echoes. Another group of echolocating birds it the cave swiftlets of South East Asia *Aerodramus* spp. The nests of these birds are used to make bird's nest soup, and are found in very dark caves. Swiftlets are diurnal, hunting for insects by day and returning to the caves at night. Their echolocation capability apparently allow them to stay out long after sunset, and to set out before sunrise, giving the colony a chance to exploit a larger area of food reserves than would be possible if they were using vision alone.

Some species of swiftlets emit impulse bursts, like the oil birds, while others give pairs of clicks, similar to those of *Rusettus* fruit bats except that they are in the human hearing range. One species, *A. vanikorensis*, was found by Donald Griffin and Rod Suthers of Rockefeller University, to produce clicks between 4,500 Hz and 7,500 Hz. With these less detailed picture of their environment than can bats Nevertheless, in tests *A. vanikorensis* was able to avoid 6 mm diameter wires in a darkened experimental room. It is not known whether echolocation clicks also serve a normal, intra-specific, bird-to-bird communication function.

**'Boomers'**

Some birds produce very low frequency calls that can travel for long distances. The booming 500 Hz note of the bittern *Botaurus stellaris*, for example, can be heard several kilometers form the reed beds. In on case it was reported to have been heard five kilometers away. The male American sage grouse *Centrocercus urophasianus* has an inflatable throat pouch with which it amplifies its booming leking call. The lek is an arena in which a group of male animals compete with each other for mating stations. Females visit the lek when ready to mate and are attracted most to the males occupying the central sites. Another bird which researches felt ought to be a 'boomer' but which, at first hearing didn't appear to be is the capercaillie *Tetrao urogallus.*

The cock bird visits the lek, fans out its tail, raises its head into the air and bursts forth not with a deep boom, bur with a rather insignificant series of clicks and pops. To Robert Moss, at the institute of Terrestrial Ecology, Banchory, this seemed rather odd. The cock obviously puts a great deal of energy into the call and can be seen shaking all over. It inflates the enlarged oesophagus like a balloon, as much as other grouse species do. With this apparatus the sage grouse, for example, can be heard at 200 meters. Moss reasoned that. As the voice of the cock grouse is much lower than the hen's, so too should be the cock capercaillie's. The hen capercaillie has a deep voice, the cock's should be deeper, perhaps infrasonic Moss went on to test this. Together with Ivor Lockie, of the Robert Gordon's Institute of Technology, Aberdeen, he recorded that much of the sound ids at frequencies below 40 Hz, which is inaudible to the human ear. The second problem was to explain how the cock capercaillie achieves this kind of sound production. Physics would explain that production of such a low frequency sound, on the organ pipe principle, would need an oesophagus of some great length, much longer than the size of the bird.

An anatomical investigation showed that the capercaillie overcomes this problem in a rather clever way. It has a Helmholtz resonator in its throat. This is a closed tube, and when air is blown across the mouth, in the same way that a jug-band blower plays a jug, it produces a deep booming note, which in the case of the cock capercaillie, is so deep we cannot hear it. The rare New Zealand kakapo *Strigops habroptilus* a species of nocturnal parrot which lives mostly on the ground, builds its own version of the Hollywood Bowl to project its very low frequency calls. Male kakapos gather in leks in the breeding season and boom, sometimes a thousand times an hour, for six or seven hours a night. The sound is amplified by air sacs that make them look almost spherical when calling, but this still doesn't account for the amplitude of the signal l. It was noticed that the birds call from excavated hollows , once thought to be dust baths, but research has shown that these 'bowls' are the right shape, like an outside auditorium or amphitheatre, to reflect the sound far and wide.

**'Drummers'**

Some birds are not only vocalists but also instrumentalists coots *Fulica atra* stamp their feet in the water to frighten away rivals, peacocks *Paco cristatus* accompany their flamboyant visual display with stiff feather rattling, muteswans *Cygnus olor* have whistling wing beats that are thought to keep flying birds in touch with each other, and the woodpigeon *Columbia palumbus* claps its wings together. But probably the most obvious instrumentalists are the woodpeckers, which drum on tree trunks with strong beaks. Dieter Wallschlager, of Humbolt University, East Germany, has been studying the woodpeckers drumming behaviour and has found that, although birds drum on different trees and produce different quality notes, it is the temporal pattern which is important. Some birds drum fast, others slow, and yet others speed up during their performance. The great spotted woodpecker *Dendrocopos major* drums accelerando, whereas the lesser spotted woodpecker *D. minor* drums with constant time intervals, although these vary from winter, when it drums slowly at ten to 12 beats per second, to spring when the rolls are faster.

The black woodpecker *Dryocopos martius* of central and northern Europe, incidentally, produces a drum roll of 40 beats, which lasts for a couple of seconds and black woodpeckers in different geographical regions have drum rolls of different durations–distinct drumming dialects. Drumming appears to be important during courtship, to bring the female into the breeding condition , and for territorial proclamations. Playback sounds of drumming will elicit drumming behaviour, for instance, in male great spotted woodpeckers. Wallschlager's work has also revealed that contained in the drumming pattern, are factors which show individual identity. Each woodpecker has its own drum pattern, so a male and female can instantly recongnise one another, as well as their neighbours, by sound. Drumming is thought to have evolved as a byproduct of feeding behaviour. Woodpecker attack the trunk of the tree for the beetle larvae and the like, hammering their beaks into cracks and crevices. Green woodpeckers *Picus viridis*, unlike their cousins, do not drum so much, but give a loud laugh-like call known as the 'yaffle'.

## Bird Songs

Usually it is the male bird that sings. There are few exceptions, but in about 95% of cases the male is the solo songster. The male only sings at certain times of the year. In Europe, most song birds sing in spring. The song is often more complicated than a call, and naturalists over the year have associated this more complex signal with the transmission of a more complicated message. The non-evolutionists, non-selectionists or fundamentalists have suggested that bird own pleasure, or indeed, as some have proposed, for the pleasure of listening ornithologists. Romantic as this may seem, there is unfortunately no evidence that birds produce their songs for the sheer joy of it. A song is difficult to produce, expensive in energy terms (as energy consuming as flying, according to some estimates) and can be very dangerous in exposing the individual to predators. A small wren, for example, belts out a very loud song for its size; the entire body snakes and quivers.

In nature, it is unlikely that a bird would make such an effort, consume so much energy which requires so much food, if it did not have a good reason for doing so. At present there is not clear evidence that bird songs contain more complex messages than calls. There appears to be no complicated code or language which we might unravel. Bird song, simply, conveys the information more effectively; the message is getting across better. One feature that is immediately obvious is repetition; the same syllables and notes are repeated in the same order and in the same pattern time and time again. With a background of the songs and calls of the other animals, atmospheric noises, and other extraneous sounds in the environment, a bird might find it difficult to get its message through. By repeating its signal a bird will eventually get its message to the intended recipient, with less risk of it being lost in the transmission process.

Male birds sing either to repel others male or to attracts females. Charles Darwin proposed that a male song bird's main interest is in charming a female to approach, share its territory, and be its mate, and that mate attraction is the principal function of bird song. Eliot Howard, on the other hand, writing in *Territory in Bird Life* in 1920, convinced most naturalists that bird song is mainly concerned with staking out defending a territory. Which, then, is the more fundamental function of song? It could be that the same song has two different meanings, depending on which individual is receiving it. A resident male might sing to repel an intruding *male* that enters its territory, yet sing the same song to attract a passing *female.*

It is not uncommon for song to have this duality. Any sophisticated communication signal is coloured by the context in which it is given and received, so the way these functions emerge depends on the life cycle of the bird. In the temperature regions of Europe or North America the male birds tend to arrive at the breeding areas first. They go through a phase of intensive competition for territories but, after two weeks at the most, they will have staked their claims and be sitting pretty on the basis of whatever agreement has been reached with neighbours. They don't stop singing continues, but to attract and stimulate females. So, even though the same song may be used for both territorial proclamation and courtship, there is often a separation in time. Not all birds, though show this duality of song function. Song may be weighted more towards one than the other. Peter Marler has suggested there is a continuum of species, from those in which mate attraction is the more important function, through to those which are more concerned with the repulsion of rivals, with all shades of grey in between.

If the sole aim is to attract a female, a male's song should be varied, with no breaks, and terminated after a successful mating. When repelling rival males song can be simple and should be spaced out in order to listen for replies. Charles Hartshorne, in 1956, described some birds as 'continuous' singers–males produce a stream of song with no gaps–and others as 'discontinuous' singers–they sing for a few seconds, pause, sing again and so on. Hartshorne, originally, interpreted this finding in his monotony threshold hypothesis–put simply, the more continuously a bird sang, and the greater the repertoire, the less 'bored' a listener would be. More recently, Peter Slater, of the University of Sussex, has pointed out that, if a song is to be used in a two-way conversation, such as that between territorial males, then there have to be gaps in the performance so that the each may hear the other's reply. 'Discontinuous' singers, slater argues, are likely to be species that use songs in territorial interactions between rival males. They would tend copy very accurately in order to match their neighbours, for their languages has to be shared. 'Discontinuous' singing in addition, has the advantage that gaps between or within songs allow the bird to keep an eye and ear open for predators, especially as it is drawing so much attention to itself in a vocal border dispute.

'Continuous' singers do not wait for a reply. They are interested mainly in attracting a mate. Continuous singing is likely to result in more improvisation and less copying. The song is simply a continuous announcement that an unmated male is here, ready, and waiting. The great tit, chaffinch and white crowned sparrow *Zonotrichia leucophrys* are extreme examples of discontinuous' singers and the nightingale *Luscinia megarhynchos,* sedge warbler *Acrocephalus schoenobaenus,* grasshopper warbler *Lacustella naevis* and brown thrasher *Toxostoma rufum,* are 'continuous' singers. Song may have other functions.

It is important that the receptive female is raised to the some state of readiness to mate as the male if mating is to be successful and fertilization to take place. At Occidental college, Luis Baptista and Martin Morton carried out tests to see if the singing of male white-crowned sparrows in any way influenced the physiology of the female. During the winter, Baptista and Morton created artificial spring and summer conditions in the laboratory. Because it was winter and out of the normal breeding season, the females were not physiologically ready to mate. In the artificial spring with extended daylength, however, the female ovaries grew. In one group of birds tape recording of males were played and the ovarian growth rates were considerably higher than in those that heard nothing.

In some colonial birds, courtship songs and calls may bring an entire colony into synchrony the advantage being that all eggs and young will arrive and grow up together; there is safety in numbers. Some birds sing 'prettier' song than others. The male blackbird, for example, sings and elaborate song with a variety of song types, repetition of phrase and various on song themes. But other birds can achieve the same results with a much less complicated signal. So, why the elaboration? Is it that the basic information, which repels male rivals and attracts females, is transmitted more effectively in a complicated song, or is there more information contained within a signal than we have so far been able to identify?

Clearly, redundancy, that is the repeated transmission of the same signal, will get the message trough, eventually, to the listener, despite background noise and other environmental constraints. But that still does not account for the beautiful complexity of the melody of a blackbirds song. To look for an answer we need to turn to Darwin and evolutionary theory. If the ultimate function of bird song, whether through territory acquisition or courtship, is to facilitate successful breeding, then sexual selection would operate most strongly when there is intensive competition for mates. Song elaboration may be an example for this kind of competition. Donald Kroodsma, of the University of Massachusetts at Amherst, wanted to know why some male birds have more then one song type. He took 24 unmated, virgin canaries and divided them into two groups. To one group he played the normal complex songs with 35 song types. To the other he played artificially simplified songs containing few as five song types. It turned out that the female canaries exposed to the more complex songs were turned on physiologically to breed more quickly that were the females that heard the simple songs. He was able to monitor this by using a technique developed by Robert Hinde at Cambridge.

Bundles of ten centimeters long string were placed in the cages, and each female would place the strings into nest cup as if building a nest. Kroodsma simple had to count daily the number of strings being used. Those hearing the complex songs built their nests faster. They also tended to lay more eggs, although without a male present the eggs were infertile. This suggested that makes with larger repertories, within a species, might have some kind of advantage when the females choose their mates. If the male can 'impress' the female with a more complex song and larger repertoire, then it might say something about the quality of his breeding potential. Form the work of Fernando Nottebohm it is know that male canaries with larger song repertories have larger song control centers in the brain, For a male song bird to learn and develop a large song repertoire must mean a considerable investment. In the canary, males rend to learn more songs and develop larger repertories as they get older. May be the female can spot the correlation between the number of different songs and the age of the male, and choose the older, more mature and successful bird. It is known in, for instance, the redwinged blackbird *Agelaius phoeniceus* that the older birds are the more successful breeders. The same is true of some other species of birds and of frogs toads.

In another study, this time in the wild and with Bewick's wrens *Thyromanes bewickii,* Donald Kroodsma followed the movements and activities of birds identified by coloured bands on their legs. The wren population in Oregon sings throughout the years. He sat, watched, and recorded the songs of seven individuals, from one or two o' clock in the morning until way after dawn, usually until about 11 o' clock. He found that the numbers of songs each male songs was plotted against the date it had hatched the previous year there was an interesting correlation; the earlier the male had hatched, the larger was the number of songs each male song in its repertoire the following year. Kroodsma suggested that the female can recognise this. After all, a male hatched earlier in the year might have a better territory would give the female a better chance of reproducing successfully. Kroodsma emphasizes that the reasoning is a little contorted and requires several inferential leaps. However, it seems likely that the sooner a male starts learning, gains a territory, attracts a mate, and begins to breed the greater chance he has of gaining an advantage over his rivals, and this may be associated with the quality of his song.

For some species, therefore, a larger repertoire appears to be desirable. By having rich and varied repertoire, a bird is thought to gain some evolutionary advantage. It came, then, as a surprise for Donald Kroodsma, in collaboration with Rick Kennedy at Rockefeller University, to find that birds of the same species in different in the songs of winter wrens, *Troglodytes troglodytes hiemalis*. In Oregon, males may have up to 30 different songs, while in New York and around Massachusetts they have only two. The European birds have three or four different songs. The same thing has been observed in long-billed marsh wrens *Cistothorus palustris*. Western males have an average of 150 different songs, eastern males have about 50. This is correlated with a reduction in the numbers of song control centers in the brain of the eastern birds. The question with which the researches are left is whether the difference is environmentally induced or genetic. Kroodsma awaits the results of new experiments.

Whether 'continuous' or 'discontinuous', repulsive attractive or stimulating, a male song bird must know when to stop. What tiggers a bird to sing, and how does it know when to terminate its song? If a male bird is isolated in a sound proof chambers so that he can hear nothing other than the sound of this own movements, he will sing to a regular schedule. He will commerce singing early in the morning as if in the wild and part of the dawn chorus. There is an endogenous circadian rhythm, an internal clock, that governs the times he sings and guides his singing programme though his repertoire of songs. Superimposed on that schedule, in the wild, are other stimuli that might trigger song production. If a rival male starts up, for instance, then a male will sing in reply.

Inhibition of singing might well result from the sudden arrival of a female. The male, in this case over to another set of vocalisations which are involved in close-range courtship. In some birds, at that moment, song virtually ceases for the rest of the season in certain to be one of the few has failed to obtain a mate.

**The Dawn Chorus**

At the break of day and at dusk, many woodland birds indulge in long bouts of singing and countersinging. To rise with the lark' or at 'cockcrow', we often says; but why should the lark and the cock be calling at that hour? Singing, as just one of a bird's daily activities, is clearly competing in a bird's schedule with foraging for food, or keeping watch for predators. Why, then, should singing be the dominant activity during the early part of the morning? To find an answer to this, John Krebs and Alex Kacelnik, of Oxford University, looked at the way great tits apportion their time to different activities throughout the day. They reasoned that early in the morning, When the bird makes up, the day is dark and cold and not a particularly good time for food gathering. Food availability is generally fairly low.

The insects, on which great tits feed, would be inactive and difficult to locate be low. John Krebs had also noticed , from his field observations, that many birds are actively seeking living space early in the day, so there would be high pressure on a territory holder from wandering birds. Both these factors, if they could be shown to operate, would encourage birds to sing at dawn. To test this hypothesis, Alex Kacelnik set up large aviaries and introduced male great tits who promptly established territories. Kacelnik was then able several factors-for example, the availability of food and the pressure from rivals. A third factor he looked at was the bird's internal clock. Might a daily clock, independent of food or rivals, govern when the birds sings or when to feed? By looking at these three factors, Kacelnik was able to establish that food was the major influence determining whether a bird sings at dawn or not.

A bird presented with an abundance of food early in the morning is less likely to give up feeding in order to sing at, and chase away, a rival. If the food is reduced in quantity there is a greater chance the bird will sing. Kacelnik also measured the efficiency with which birds feed at low light intensities and low temperature and confirmed what Krebs had intuitively expected; that given the temperatures and light intensities normally found at dawn, birds are less good at feeding than, say, three hours after dawn. With atmospheric conditions unfavourable for foraging, there is likely to be greater pressure from territory-seeking males. Dawn, therefore, is the part os the day when the greatest benefit is derived from chasing away. Acoustic theoreticians had already shown that transmission are more effective at dawn because air turbulence is at a minimum and temperature gradients are favourable. But the work of Krebs and Kacelnik revealed that the picture is far more complex that was first thought, and food availability and pressure from intruders play a major part in any explanation of why birds sing at dawn.

**Territorial Proclamation**

Song can be long range 'keep out' signal, a proclamation of ownership, but how do we know that there are not other forms of signal that are just as important? In order to check this, John Krebs went back to square one. As a population ecologist, his interest in bird song came from the population dynamics of birds for his thesis, Krebs was investigating whether the breeding density of birds in a particular habitat is limited by territorial behaviour. First he wanted to be sure that intruders are kept out by some kind of behaviour on the part of the territory holder. One spring, he carried out some simple experiments which involved removing territory holders from a breeding population of great tits. He found that new birds would enter the vacated territories, filling up the empty spaces.

The new arrivals, he noticed, came from poor habitats nearby, where breeding success was low. They had been waiting on the sidelines, ready to move house when a better living space became available. How do the birds know whether a territory is occupied or not? The intriguing thing was that the birds would come in very quickly after the removal of the territory holder. The obvious answer seemed to be that they listened to see if a resident was singing. Sound signals would be more effective in a forest than say, visual or olfactory cues. Krebs then checked the literature for references to song and territorial proclamation. There were lots of studies showing that, when song was played back to territory holder, it would react by approaching and may be attacking the loudspeaker, and this was taken as evidence that song acts as a territorial signal. But no-one had checked that, when a bird sings in its territory, it is actually saying 'keep out' and therefore defending its territory against the birds waiting in the wings.

In 1975 Krebs carried out his classic, yet simple, removal experiments. He took away territory holders, as before , leaving empty spaces in the wood. This time he replaced some of them with loudspeakers playing the songs of the birds he'd taken away. In other territories he put loudspeakers playing a control sound–a tune played on a tin-whistle, which had notes at about the right frequency and duration. Other territories were left empty. He sat and watched. Sure enough, in came the birds from outside the study area to occupy the empty spaces. The territories with normal song were treated as if a bird was in residence and were avoided. Instead, the new arrivals went first to the tin-whistle or silent spaces. This seemed to show that the initial idea was correct–male great tits waiting to enter and take over territories know whether a territory is occupied or not simply by listening for the resident male's song.

Having established a territory, a bird must maintain and defend its borders. In some species, neighbours sing at each other across the line of demarcation. Often they sing at same song, a form of countersinging known as 'matching', But which bird should sing first? And, as most birds have more than one song type, how does an individual, having started, know which song type to sing next? Donald Kroodsma has uncovered some clues while working with the long-billed marsh wren, a bird with a large, rapidly sung, song repertoire. He was fascinated by the way males in neighbour territories sing the same song repertoire back and forth to one another each day. Bird species with very simple songs would have little scope for improvisation when matching with neighbours. But with a couple sing together is limitless. How do they decide the order in which to sing? In a pilot project Kroodsma taught two wrens nine song types. They were allowed to sing at each other quite freely and it soon became apparent that the physically larger male became dominant, and led the singing order. Kroodsma then interfered with the singing.

The dominant bird began to follow the amplified bird's song and the previously submissive bird started to lead the proceedings.

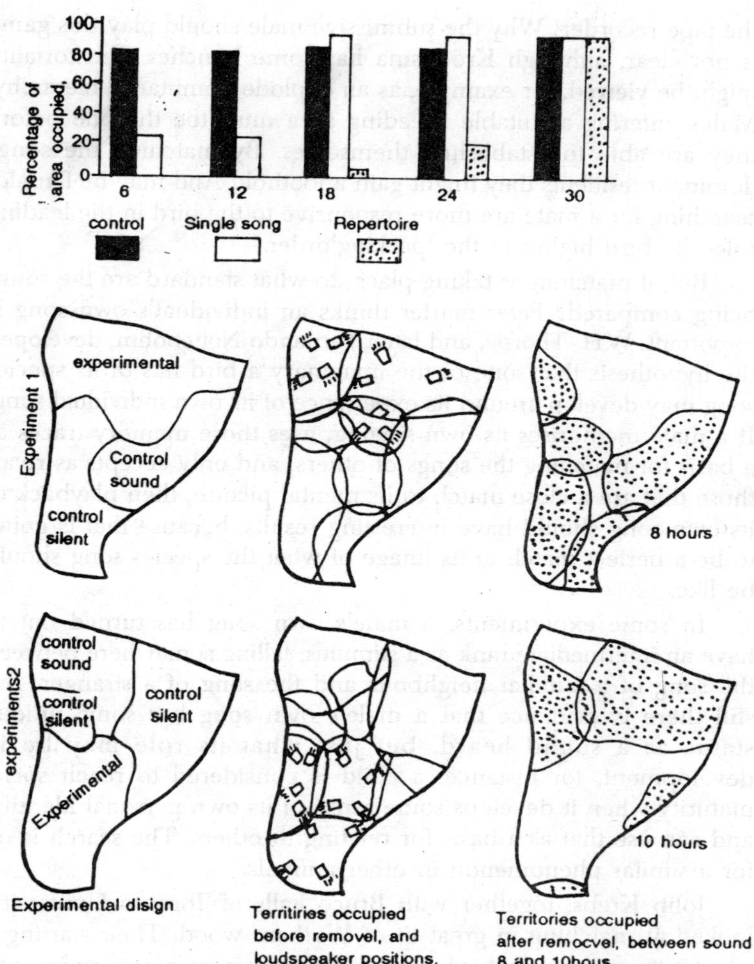


*Fig. 7.2. Removal and playback experiments with great tit songs. Resident males removed and loudspeakers placed in certain territories.*

In the field, Jerrard Verner studied eastern Washington long-billed marsh wrens.. He Placed a loudspeaker playing marsh wren songs in an established territory to stimulate an intruding male. Normally a resident is dominant to an intruder, and verner found that, because the order of singing in marsh wrens is so stereotyped, the resident bird was able to anticipate the order of songs on the tape recordings and get in with the matched songs first. In this way it actually led the tape recorder! Why the submissive male should play this game is not clear, although Kroodsma has some hunches. Territoriality might be viewed, for example, as an explode dominance hierarchy. Males entering a suitable breeding area must toe the line before they are able to established themselves. By matching the songs dominant residents they might gain a foothold And may be females searching for a mate are more responsive to the bird in the leading role–the bird higher in the 'pecking'order.

But, if matching is taking place, to what standard are the songs being compared? Peter marler thinks an individual's own song is important. W.H. Thorpe, and latter Fernando Nottebohm, developed the hypothesis that sone of the imaginary a bird has of its species song may develop around its experience of its own individual song. If a male memorises its own singing, uses those memory traces as a basis for matching the songs of others, and only accepts as rivals those that are a close match to its mental picture, then playback of its own song should have interesting results, because that is going to be a perfect match to its image of what the species song should be like.

In some experiments, a male's own song has turned out to have an intermediate rank as a stimulus, falling somewhere between the song of a familiar neighbour and the song of a stranger. But still there is evidence that a male's own song has some unique status as a sound heard, but just what its role may be in development, for instance, a child is considered to reach social maturity when it develops some sense of its own personal identify, and can use that as a basis for relating to others. The search is on for a similar phenomenon in other animals.

John Krebs, together with Bruce Falls of Toronto University, looked at matching in great tit of Wytham wood. Their starting a point was to play to birds songs from their own repertoire and record whether or not they were matched. They found, not surprisingly, that great tits match their own songs. What they also found in that the reliability with which they matched varied according to circumstances, but it was not clear what those circumstances were, They looked in more detail. In another experiment birds were presented with three different kinds of playback of song types from their own repertoires : firstly their own version of that song type; secondly a neighbours version; and thirdly a version of the same song type sung by a stranger.

Typically, birds would approach the loudspeaker, fly about looking for an opponent, and then begin countersinging. The intensity of the challenge was not always the same though. A neighbours version of the song resulted in a week response. The resident bird would approach slowly and sing little. A strangers version would evoke a full bodied challenge which sometimes ended with the resident attacking the loudspeaker.

Bruce Falls feels that this may be because the stranger poses a greater threat than does the neighbour. The neighbour has a territory of its own and is recognised the amount of unnecessary effort, time and risk spent in fighting and chasing, by ignoring, to a certain extent, the song of immediate neighbours. The accuracy of matching varied too. Birds most reliably matched their own songs; next most reliably matched the songs of neighbours, and least reliably matched the songs of strangers. This still did not tell what purpose matching serves; rather it showed that matching is based on a precise similarity between the song version played to the bird and the bird's own rendition of that particularly song-type. Krebs looked at this data again and came up with a possible functional significance of matching.

Krebs remembered the environmental degradation experiments with Carolina wrens carried out by Haven Wiley and Douglas Richards. They had revealed that Carolina wens are able to judge the distance to rival bird, which is singing, simply by listening to the way the signal is degraded by echoes, reverberation and other atmospheric disturbances. But how, wondered Krebs, can a bird actually tell if a song is degraded? It would need to compare the song it hears with a standard of what the song would be like if it were not degraded. Krebs concluded that the standard is in its own repertoire. When a resident hears a song, it is compared with its own rendition of the same song and the bird can assess whether the song is degraded or normal, thus discovering the distance of a rival. So if two neighbours match each other's song types, the receiver can say 'I have the same song type as you, therefore I can judge exactly how far away you are, because I can measure the degradation of yours song', and the original singer is telling how fat away it is. It is a mutual exchange of information about distance, and reliable too. The degradation depends solely on the properties of the habitat, pretending to be closer or further away than they really are.

By matching to its own song, a bird can tell exactly whether a neighbour is say, 25 meters away. The five meters difference could be important because a neighbour 25 meters away might be inside its own territory and no threat, but 20 meters away it might be inside the other's patch and attempting to steal a piece of ground. Neighbours save time and energy if they keep telling each other how far apart they are. If the song a bird hears is very different from its own version it would be difficult to judge easily the amount of degradation; hence the more violent or alarmed reactions noted with intruding strangers. Repertoire size was another topic which interested Krebs.

The well-known arguments to which he turned his attention was that if bird song is saying something fairly simply like 'keep out' this is my territory', or 'here I am an unmated male', why is it said in such an elaborate and complicates way? Krebs went back to his removal experiments, taking male great tits out of their territories and replacing them with loudspeakers. The playback experiments were repeated but this time, different numbers of songs were played in the territories with loudspeakers. In one territory a varied repertoire of song type was played, while in another a single song does play a role as a 'keep out' signal, then perhaps the repertoire, in some way, increases the effectiveness of song in repelling rivals. When the residents were removed and the tapes played back, new birds would arrive and scout around to find the empty spaces. The spaces with no sounds were quickly reoccupied as before, but interestingly the spaces playing the single repetitive song type were more prone to occupation than those with the varied repertoire, it seems, make better' keep out' signals; but why?

John Krebs feels the mechanism is a very simple biological property of animal nervous systems. When exposed to a repetitive stimulus an animal often habituates, that is it stops responding to the stimulus. If a territory holder sings the same song over and over again, rivals might gradually get used to it. If the resident bird changes its tune, it variety in its song reduces the chance that others will habituate to it. Why wandering birds should habituate to song is not at all clear. May be it is simply a property of their brains. Krebs would prefer to find a functional explanation, but as yet is not one.

Territory holders with varied repertoire have an opportunity to cheat. Wandering birds will be on the lookout for suitable areas with few resident birds will be on the lookout for suitable areas with few resident birds, where there is less competition for mates and food resources. By singing a variety of songs, a resident bird might give the false impression of there being many other birds in the area. The deceit could be further enhanced by constantly moving about the territory singing from song-post to song-post. It would seem more profitable then, for an intruder to look elsewhere. This has been named the Beau Geste Hypothesis after the novel by P.C. Wren in which the gallant legionnaire successfully defends a desert fort by propping up dead around the ramparts to make it look as though there were many more defenders.

One bird which is seen to dash about its territory incessantly often to the irritation of the wildlife sound recordist–is Cetti's warbeler *Cettia cetti.* It may sing its first song in front of you, but it will disappear for a few seconds only to reappear at another song-post, a little further away, to sing its next song. Recent research by Ken Yasukawa, while at the Rockefeller University Field station at Millbrook, with the red-winged blackbird of North America has given some credence to the Beau Geste Hypothesis Analysis of song length and the frequency with which songs are sung has shown that they are as variable within a single bird's repertoire as they are between repertoire of a group of different birds. Also male red-winged blackbirds tend to change song types when they changes song-posts.

So, come foul means or fair, if a song repertoire is important in obtaining the best territories, there should be a link between repertoire size and reproductive success. Male great tits, for instance, breeding in the choice territories should have mates laying the largest clutches of eggs and therefore producing the greatest number of surviving young to breed the following year. In Wytham wood, John Krebs, Chris Perrins and Peter MacGregor recorded the songs repertoire with the ability to rear successful families. As part and his assistants go around the wood and count the number of eggs laid by each pair of birds, how many hatch out to produce young, and the number of young that survive through the winter to breed the next season. Peter MacGregor and John Krebs could thus record the songs of known individual males.

The Oxford researches showed that there was, indeed, a relationship between the size of song repertoire and breeding success, although the result was not the one expected. Males with poor repertoire containing one or two song types, and males with large repertoire of seven or eight did not fare so well as those with middle-sized repertoire of three to four song type. There seems to be, in great tits at least, an advantage in having too many. Males with poor repertoire do worse in terms of reproductive success than males with several song variants do have enhancing effects on the territorial keep-out signal, which are reflected in reproductive success, he is at a loss to explain the disadvantage of very large repertoires.

## Courtship

Human observes are often attracted by the conspicuous and repetitive movements and the loud elaborate sounds made by some birds prior to mating. Courtship is an important aspect of bird behaviour, for its creates the right conditions in which the male and female of the species can come together in order to reproduce. For birds, visual and acoustic signals are used as important sexual attractants. For birds in marsh or forest, where visibility is restricted, songs and calls are often the only means of long-distance communication. In song birds, it is the male that sings. Instead of pursuing a female, a male sedge warbler, for example, sits and sings in his territory and waits for the female to come of her own accord, attracted by his song. The problem is the sound might not only attract a mate but also a predator.

A courtship display is a compromise between attracting a member of the opposite sex and exposure to a predator. And, to complicate things still further. to avoid wasted effort, courtship singing must attract only females of the same species. Song must be recognisably distinct for each species. A bird is also likely to be agitated about the close proximity of another individual of the same species. The potential mate could be confused with a marauding intruder or a competitor for food. A male's song, therefore, must not only attract the female, it must also suppress her natural tendency to flee. From the female's point of view, not all males are equally suitable for mating. She must have a means of selecting the fittest. The strongest and Healthiest males usually have the highest reproductive potential–they will look after the brood better, and have the best territory with the most food. These are short-term benefits. In the long term the male offspring are likely to inherit the father's ability to maintain and defend the best available territories, so there is an obvious genetic advantage in selecting the best available male.

A male's song may indicate to a female his suitability as a mate. So could it be that the ultimate function of bird song is sexual attraction? Clive Catchpole, of Bedford college, university of London, thinks this may be so. He has been working with several species of warblers, including the reed warbler. *Acrocephalus warblers A. schoenobaenus* which breed commonly in Britain, and the marsh warbler *A. palustris* which is largely confined in Britain to the south-west of England. The *Acrocephalus* warblers are small, well camouflaged birds which live in dense marshland vegetation. As they cannot see each other clearly there has been an obvious premium on vocal, rather than visual communication in their evolution. The first clue Catchpole was able to uncover that song might be particularly important in sexual attraction for these species came from studies on their seasonal and daily rhythms of song production. They are migratory birds, the males arriving in Europe during the spring to take up sometimes throughout the day and night. The females arrive a few days later, find a mate, and settle in for the breeding season.

After pair formation the male's routine changes-he reduces song production, and in the case warbler, stops singing altogether. This indicated to Clive Catchpole that the males is using his song to attract a female. Once he has found her there is no need to sing, so he stops. Catchpole was able to confirm this with playback experiments. He placed a loudspeaker in a sedge warbler's territory and played a rival's song. The resident would approach the source of the sound aggressively, hunt around, fluff-up in a threat display, but he would not sing. He would still defend his territory, but not with song. Further evidence for sexual attraction, suggests Catchpole, lies in the warbler's elaborate song. The sedge warbler can have a hundred or more separate syllables. The male selects form his repertoire several syllables which are composed into a long and elaborate song by repeating them, alternating them, and recombine them in a variety of patterns. The patterns are so variables that Catchpole, having analysed miles of recorded tape, has been unable to find an instance of a sedge warbler repeating a song. Each song is a separate musical composition, a distinctive behavioural event. Catchpole likens the sedge warblers song to the peacock's tail in that it is a flamboyant and extravagant signal used for impressing and attracting females.

To test whether male sedge warblers singing complicated songs have greater success in attracting females than do those, with simpler songs. Catchpole recorded a wild population of sedge warblers in England. He noted when each bird arrived, and if and when each one attracted a mate. This was relatively easy to do because mated sedge warblers stop singing. So, the date of arrival and the date on which the male attracted a female to his territory was field away with tape recording of his song. The songs were then analysed with the sound spectrograph and the complexity of each rendition assessed. The number of elements in each bird's repertoire was plotted against the pairing date and it was revealed that the birds with most complicated songs attracted females first. Those with simple song repertoire had to wait. Birds with elaborate songs clearly had some advantage over their rivals. Not all species of warblers produce such complex songs as the sedge warblers, and yet others sing with even more intricate song patterns. By looking at the species in his study group, Clive Catchpole sought some clue to the evolution of the songs. He looked at their ecology and behaviour and found major differences.

Some *Acrocephalus* warblers are monogamous–one female per male–and others are polygamous–each male attracting several females. Most bird species are monogamous. Both parents are needed to bring food to the nestlings or the chicks may starve. However, in certain habitats food can be plentiful. In marshland, for example, insects life can be super-abundant. In areas rich with food supplies, both parents may not be needed to feed the young. A female could catch enough large insects near the nest to bring up the youngsters by herself. The male, in that circumstances, might desert the female confident that his genetic investment is safe, and polygamy evolves in the *Acrocephalus* warblers, polygamy has evolved in two European species the aquatic warblers. *A. paludicola* and the great reed warbler *A. arundinaceus.*

Clive Catchpole first reasoned that, if sexual selection has driven the evolution of elaborate bird song, then polygamous males will have evolved more elaborate songs. A 'showy' male, attractive to a bevy of females, would leave behind more offspring. Catchpole tested the prediction by assessing the song syllable repertoire of each species. He took the results from the six European warblers, but was in for a surprise. It was, in fact, the polygamous species, the aquatic and great reed warbler, that had a song repertoire of only ten to 20 syllables, and the monogamous species, the reed warbler, the sedge warbler, the marsh warbler, and the moustached warbler. *A. melanopogon* that had the most complicated songs with 35-100 song syllables. The polygamous bird sang short simple songs while the monogamous males produced long, continuous and variable bouts of song. For an explanation Catchpole looked at the bird's ecology and behaviour.

| *Species* | *Repertoire range: Number of syllables* | *Mean Length of song in seconds* | *Mating system* |
|---|---|---|---|
| Marsh warbler (A.palustris) | 80-100 | Continuous | Monogamous |
| Reed warbler (A.scirpaceus) | 70-90 | Continuous | Monogamous |
| Moustached warbler *(A.melanopogen)* | 60-80 | Continuous | Monogamous |
| Sedge warbler *(A.schoenobaenus)* | 35-55 | 19.49 | Monogamous |
| Great reed warbler *(A.arudinaceus)* | 10-20 | 3.20 | Polygynous |
| Aquatic warbler (A.paludicola) | 10-20 | 2.18 | Polygynous |

It seemed to catchpole that, in a polygamous species, a female would less concerned with the quality of the male, as she is likely to be deserted and left to bring up the chicks herself. To maximise her reproductive success she should be more concerned with territory and the food it will provide for her young. The female is not selecting her mate directly on his own attributes, but indirectly through the quality of the territory. Polygamous species defend large territories in rich marshland. So male song, Catchpole believes, may have evolved primarily within the context of competition between males for territories–intra-sexual-selection pressure–with the resulting short, simple songs. This further supports Peter Slater's ideas on 'continuous' and discontinuous' singers.

In monogamous species, with long elaborate songs, the male is important in helping to feed the young. The female chooses a good quality male rather than a good quality territory. Territories in monogamous species of *Acrocephalus* tend to be small, sometime in drier areas, with most food being obtained from way outside the territorial boundaries. The birds make long flights in search of food supplies so the male must be strong, physically capable of many arduous flights for foraging. The first indication a female may have of a male's suitability is his song which she hears some distance away. She may, or may nor be attracted by what she hears. This is inter-sexual selection at work. As in the sedge warbler study, the males with the most elaborate songs obtain their mates first. There may in fact be a mutual advantage–the male, by singing an elaborate song, reaches the more active and alert female, while the early female gets to choose the cream of the males. Clive Catchpole believes that the ultimate function of bird song is to obtain a female, breed successfully and produce more offspring than other individuals. Even those birds singing to defend territories are selected for through the quality of the living space. The end results is that a male obtains a female by the indirect route of proclaiming a territory.

Catchpole also points out that the repulsion of males and the attraction of females are not mutually exclusive. A song may serve both functions, although in different proportions. In the sedge warbler, which stops singing after pairing, it is clearly mate attraction that is more important. In other species, which reduce their song output after pairing, but retain it for use later, it may well be that there is some territorial function versions of its song. It is a simple song, but with a certain amount of variation. It starts with a number of short clicking sounds and expands into more melodious notes towards the end. When the birds is setting up a territory and trying to attract a mate is sings the longer, more elaborate version of the song, each song lasting from between three to 20 seconds. When a female has been attracted, the male stops singing the long complicated song and instead sings the clicking sounds. Catchpole confirmed this with playback experiments. He placed a loudspeaker in the territories of birds which were singing the long more elaborate song. If he played back the song of a rival, the resident bird would approach the loudspeaker, at the same time shortening his songs–the longer, complicated version used to attract a female, and the shorter, simpler, more stereotyped version it continues it continues to use to keep out rival mates.

# 8

# Behaviour in Bird Species

In birds many behaviour are inherited and thus or *"instinctive"* – they are performed without preliminary experience or learning. Innate behaviours normally occur in patterns–*fixed action patterns*– so remarkably constant that they must be triggered by stimuli can be discharged only after being activated by *internal releasing mechanisms* selectively responsive to the stimuli. The stimuli triggering a particular innate behaviour are special features of the environment–from, colour patterns, movements, sounds etc.–that function as cues and are called *releasers.* Practically all bird species have social releasers, frequently termed *signals* or *displays.* These may be visible or audible, and their principal and often exclusive function is to broadcast information to fellow members of the species. Two near-classic examples of social releasers, proved by experiments, are the black malar strip (*mustache*) of the male Yellow-shafted Flicker (*Colaptes auratus*) that releases courtship behaviour in the female and attack by an intruding male and the red patch on the lower mandible of the Herring Gull (*Larus argentatus*) that releases food-begging in the chick. Both examples are derived from markings that stand out in contrast to the adjacent colour and are thus conspicuous. Other examples of releasers that send out stimuli are particular movements and positioning of the head and/or other parts of bird's body, vocalizations such as songs and alarm cells, and so on.

Through experimentation, *Tinbergen* discovered that he could produce *super-normal stimuli* that would elicit a more effective response. For instance, he found that an Eurasian Oystercatcher (*Haematopus ostralegus*) which normally lays three, sometimes four eggs, prefers five if given a choice between three and five, and if offered a giant model of an egg painted to match the natural colours of its own eggs, makes frantic attempts to sit on it rather than on its own eggs. Theoretically, releasing stimuli will not trigger innate behaviours until the bird is motivated–i.e has built up a *specific action potential* through hormonal actions and other factors to perform behaviours. This build-up toward discharges of most any innate behaviour involves restlessness and exploratory maneuvers referred to as *appetitive activity*.

Once the bird is sufficiently motivated, it then responds to the appropriate releasing stimuli with *appetitive behaviour*. Provided the bird's searches are successful and the stimuli act on its internal releasing mechanisms, the build-up of action potential acquired by the bird is then discharged in the performance the *consummatory* act which consumes the energy. Two illustrations of appetitive behaviour leading to the consummatory act are in feeding and nest-building. As the motivation (*drive*) to feed or to nest develops, the bird shows appetitive activity by searching for food or for nest site and nesting materials. When, finally, by responding to stimuli, it finds and swallows the food, or it selects the nest site and builds the nest, it performs the consummatory act. For the immediate future the bird has no *desire* to eat or to build another nest. Occasionally, a bird performs an act in the absence of a discernible stimulus. This suggests some sort of build-up of motivation to the point where, in the continued absence of a proper stimulus, the energy is discharged in a behaviour called a *vacuum activity*.

A good examples is a male passerine bird bringing food to the eggs in his nest, "anticipating" the feeding of young. If the incubating female is absent at the time, he makes futile attempts to offer the food to the eggs and then departs, swallowing the food before head or taking it with him. Often a certain behaviour appears irrelevant, being "out of context". An incubating bird, disturbed and driven from its nest by an intrude, does not attack, perhaps from fear, but instead proceeds to feed when it is actually not hungry or to preen when its plumage is in ne need of attention. Two shore birds suddenly stop fighting and one of them tucks its bill under the scapulars as though it were going to sleep. These are examples of what is commonly called a *displacement activity*.

Now and then a bird directs a certain behaviour at an object to which it does not normally respond. Two Herring Gulls, in a territorial dispute, vent their hostility by pulling grass instead of fighting each other. An incubating bird, disturbed at its nest by a human being, may attack another bird, or it may peck vigorously at the branch of a tree as if in anger. Any such behaviour is called a *redirected activity*. Rather frequently a bird starts to perform a behaviour but does not complex it. For example, a bird crouches as though intending to spring into flight, but it does not fly. A behaviour of this sort, very low in intensity, is called an *intention movement*.

These behaviours–displacement activities, and intention movements which so often appear in conflict situations–are of special significance in that many seem to have become incorporated through evolution in more elaborate behaviours such as mating displays. They are then highly stereotyped or "formal" and are commonly referred to as *ritualized*. In this new role they serve as signals to elicit a response in another bird. When any one of these conflict behaviours becomes ritualized for mating, the action is more exaggerated, more emphatic, and is consequently more conspicuous. Frequently, the conspicuousness is enhanced by bright colours in the plumage involved. Displacement preening in ducks such as the Mallard (*Anas platyrhynchos*) consists of lifting sharply and spreading the secondaries and adjacent scapulars of one wing and while, turning the head under the secondaries and scapulars as if intending to preen, revealing the iridescent speculum.

Ritualized behaviour, in addition to being more conspicuous, tends to be species-specific–i.e., peculiar ti the species performing it. Thus displacement preening in the mallard differs in minor ways from that in other ducks, thereby serving as an isolating mechanism that prevents cross-breeding. *Learned behaviour* is essentially adaptive behaviour resulting from practice or experience. Basic to learned behaviour is, of course, inherited behaviour which often predetermines the extent to which learned behaviour may develop.

Learned behaviour, however, is frequently so fused with, and inseparable from, inherited behaviour that most behaviours cannot be declared the result if inheritance or learning or both. *Thrope* provided a good example from his research on the development of song in the Chaffinch (*Fringilla coelebs*). He discovered that Chaffinches reared in auditory isolation could give extremely simple songs (subsongs), not the full song of the species. The simple songs

*Fig. 8.1. A territorial dispute between herring gulls.*

were obviously inherited, but why not the full song? He found the answer when he exposed isolated Chaffinches at the age of about six months to recordings of the full Chaffinch song. They could learn to sing it. Yet other Chaffinches which he reared in isolation and exposed, at the same age, to recordings of songs of different species could not learn them well, if at all. Thorpe concluded that Chaffinches must learn the full song from their own species. It is only components of the simple songs–in a sense, the ability to sing their own songs–that they inherit. They are born with no appreciable ability to learn the songs of species.

The study of learning in birds poses numerous problems in interpretation and identification. Ceratin behaviours are sometimes misconstrued as the product of learning when learning is actually not involved. This is the case with mainly neuromuscular activities (motors activities) such as flying. A young bird just out of nest does not fly as expertly as one a little older, presumably because it has to learn how to fly. Yet a pigeon raised in a tube, without being able to use its wings, to the age when it should fly expertly, with indeed fly expertly when first released. Flying ability is inherent; it shows up not with practice or experience but with the maturation of neuromuscular coordinations.

Learned behaviours resulting from practice or experience usually develop in the early stages of a bird's life. *Thrope* (1961) found that young Chaffinches must hear the full adult song of the species during their first year of development, if they are learn it and sing it normally there after. Not hearing it, they may never learn to sing it even though they may be it, they may never learn to sing it even though they may be exposed to it later. A learned response in birds sometimes shows up to long after the process of learning has taken. Thus Chaffinches exposed to normal full songs as wild juveniles, then captured in the fall of their first year and isolated singly so that they cannot hear any other Chaffinch sing, may not show the effect of the original exposure until they begin to sing at the start of the first breeding season. Learning takes different forms, all of which are attributed to experience, not maturation, and all of which may intergrade.

*Habituation,* the declining in a response to repeated stimulation without reinforcement ("reward or punishment"), is one form of learning. Stated another way, habituation is learning to avoid commonplace stimuli that are irrelevant to the welfare of the individual. Stated still another way, habituation is becoming used to, and no longer reacting to, anything new that proves to be harmless or non-rewarding. Tameness in a bird is the result of habituation whereby the bird becomes accustomed to unnatural conditions such as a cage or the heretofore unfamiliar actions of human beings, Once the bird is tame it is likely to remain so.

Another form of learning is *trial* and *error,* the selective acquisition during appetitive behaviour of a response resulting from reinforcement. A bird learns to swallow palatable objects and to reject objects with a bitter taste. In same way, a bird when first flying learns to take off and land in the wind rather than with it and to select perches that can accommodate its feet and support its weight; a bird when first building a nest learns to select those materials that it can manipulate best. Playing (see below under Individual Behaviour) in a young bird may be a means of learning through trial and error. A third form of learning is *imprinting* which usually takes place during a brief, critical period in the developmental stage of a bird's social life. Once acquired, the learned habit may sometimes remain stable for the duration of its life. Occasionally the behaviours manifests itself or reappears long after the process of learning.

Imprinting is commonly observed in waterfowl and gallinaceous birds. Ducklings and goslings, for example, will learn to follow a substitute "parent" such as a human being, thereafter ignoring its natural parents. In the Mallard the period of "learning to follow" is between 13 and 16 hours. With age, or with the tendency to flee strange objects, the following response is apt to weaken, even disappear, though it may show up again under favourable circumstances. Imprinting is much less commonly observed in passerine and other birds reared in nests. Nevertheless, *Thrope* has suggested a form of learning, probably imprinting, in Chaffinches. The young Chaffinch, hand-reared in isolation, sings simple songs that are innate; but the young Chaffinch, reared in the wild, on in due course of time, characteristics features of the full song without apparently practicing them. Later, in its first breeding season, it learns further characteristics from competing males. After 13 months, however, it can no longer learn features of other songs because its period of learning–presumably its period of imprinting–has ceased.

There are a few species of birds which manipulate or actually use objects to solve simple problems. The Tailorbird *(Orthotomus sutorius)* of southeast Asia forms a receptacle for its cocoon-like nest by perforating the edges of broad leaves with its bill and then drawing them together with plant fibres. A number of passerine species have the ability to pull up breasted Buzzard *(Hamirostra melanosternon)* of Australia drops stones onto emu eggs in order to break them open, and from a standing position the Egyptian Vulture *(Neophron percnopterus)* tosses stones onto ostrich eggs for the same purposes. The woodpecker finch *(Cactospiza pallida)* of the Galapagos islands has the singular habit of picking up or breaking off cactus spines or twigs and using them as tools for extracting insects from holes and narrow crevices. Whether or not the bird acquires this habit by learning is open to question.

*Millikan* and *Bowman* suggest that the habit may be a form of displacement activity "when another behaviour performance has been thwarted in some way." Possibly the ability is simply an inherent behaviour re-adapted the experience to perform a task more efficiently. Only a careful study of the way the habit develops among several individuals can prove whether or not learning is involved. A few experiments have shown that birds possess quite remarkable abilities to learn, presumably by *insight*–the "sudden adaptive re-organization of experience. "For example, *Pastore* proved with a series of 21 experiments on canaries (*Serinus canarius*) that they can learn to respond to a unique object among a group of nine, the other eight being the same in all respects. For his experiments he put before each bird a board with nine small depressions in a row. In the first trial of his first experiments he capped eight of the nine depressions with aspirin tablets and the other with he hid some food, the screw, the unique object, under which he hid some food, the reward.

In the next trial the food was placed under either an aspirin tablet or a wood screw in depressions selected at random. The purpose of all the trials was to train the birds to find the food by pushing off the unique object. Each of the four canaries, after an average of nearly 160 trials, achieved the ability to make the right choice in 15 out of 20 attempts. In a subsequent series of 20 experiments, using differing pairs of objects, the average number of trials required by the canaries to choose the unique object became steadily fewer, to under 30 by the 13th in the series of *uniqueness* although they must have perceived it as abstract quality.

As another example of what may be learning by insight, *Kohler* taught a common Raven (*Corvus corax*) to "count" an "unnamed number" of objects. To do this he trained the bird to open the one box in five which had the same number of black spots on its lid as on key card on the ground in front of the box. The raven eventually learned to distinguish between five groups of spots–two, three, four, five, and six. To prove that the bird was choosing the lid with the number of spots corresponding to the numbers on the card, Kohler changed the size, shape, and arrangement of the spots on both the lids and the card in a random manner. Even the position of the boxes in relation to the keys card was changed. With this elimination of extraneous cues, the raven could still distinguish numbers of objects, but without consciously recognizing numbers as such.

## Individual Behaviour

Many behaviours are directed towards the bird's own welfare. These include *maintenance behaviours*, concerned with care and comfort of the body, and behaviours associated with selection of habitat and cover, with food and feeding and with playing.

### Maintenance Behaviours

A large numbers of movements relate to the care of the plumage and, to a lesser extent, the skin and soft parts. Since the feathers, especially those used in flight or as insulation for the body, are vital to the bird's survival, the movements directed towards their care have evolved through natural selection to an elaborate degree and become highly stereotyped. Some of the maintenance behaviours have served as fruitful sources for displacement activities and displays.

**Preening**

Probably the most important as well as the most frequent activity in plumage care; performed by the bill. Usually a bird deals with one feather at a time, seizing it at the base between the mandibles and working toward the tip, either nibbling it continuously or drawing it through the mandibles in one movement. This serves to clean the feather, to interlock barbs that have become separated, and to smooth down the plumage. Less often the bird moves the bill rapidly along the feather from tip to base near the shaft either in a stroking action or in a trembling action. When preening, the bird fluffs the feathers to make them more accessible and turns its head toward the plumage areas needing attention. Meanwhile, the bird contorts itself and assumes numerous postures, all quite stereotyped.

**Head-scratching**

Performed by the foot. Sometimes, along with preening, head-scratching is an activity for the care of the plumage, especially the feathers of the head and upper neck that are inaccessible to the bill; otherwise it is a simple reflex action in response to an irritation in the head region and is therefore a comfort movement. Birds use two methods in reaching the head: *directly*, by lifting the foot the foot straight up from under the wing; *indirectly* by bringing the foot up and lowered wing. As a rule, different taxonomic groups of birds use on method or the other. Neither is reliable as a taxonomic character since some closely related groups, even some individuals of the same species, use both methods.

**Bathing**

This activity differs in water birds and land birds. Water birds (e.g., loons, grebes, cormorants, and waterfowl), while floating immerse their heads and shoulders in a scooping motion which sends the water over their backs while they rub their sides and flanks with their heads, ruffle their feathers , and lift and splash more water over their plumage by beating their wings. Wading birds stand in the water but bathe the same way. Land birds, going into shallow water, first dip their throats and breasts and flap their wings up and down, then raise their heads and foreparts and lower their tails and hindparts into the water and flap their wings in and out of the water and over the back, spraying their plumage which is ruffled so that water can reach all the feathers and even the skin.

There are many exceptions to this method of bathing. Swifts, swallows; and other highly aerial birds, as well as hummingbirds, Kingfishers, and some Owls, drop repeatedly to the water's surface and perform a skimming action sufficient to wet their plumage. Swifts bathe in falling rain. Tropical species of many varieties bathe mainly in the rain or wet foliage. Almost all land birds will take advantage of rain or even lawn sprays. Bathing functions as a means of keeping the plumage in order, rarely, for cleaning it. While bathing, a bird is liable to predation; hence it is innately apprehensive and habitually avoids drenching itself to the extent of impeding its escape by fight.

**Oiling**

A form of preening among birds with an oil gland. Typically and often in association with preening, the bird touches or pinches the oil gland with its bill, stimulating the flow of oil to the bill's surface. The bird then smears the oil on the plumage areas with rubbing movements of the bill. But there are many variations. Long-necked birds often first rub the head of on the oil gland and then rub it immediately on the plumage areas. Short-necked birds, such as passerines, oil the head with the foot, first taking the oil with the bill, then scratching the bill with the foot.

**Sunning**

An activity among practically all birds, sometimes performed after preening, bathing, or dusting. Typically, in a moderate reaction to warm sunlight, passerine bird raises the feathers of the head, neck, back and rump, crouches with the axis of the body perpendicular to the source of light, spreads the flight feathers of the wing nearer the sun, and fans the tail. It sometimes pants with opened mouth. It may draw the nictitating membrane over the eye and perhaps close the lower lid. Non-passerine birds react in different ways. Vultures and other large diurnal birds of prey may stand, facing the sun with wings extended. Cormorants sun themselves similarly after coming out of the water or long after their plumage is dry. Birds being photographed and exposed too long under bright lamps that radiate considerable heat, give the normal sunning response. During moderate to high-intensity sunning, some birds seem to be in a semihypnotic state and much less cautions than is normal. Whether sunning directly benefits the plumage or serves any other significant function is undetermined.

**Dusting**

A behaviour in which dust–powdery dry surface soil–is forced through the plumage. In an average performance, the bird proceeds to scratch and/or peck out a depression while crouching and usually turning its body. Once the hallow is deep enough and the soil loosened, the bird sits in the hollow, raises dust through the ruffled plumage with a flicking and scooping action of the wings alternating with a scratching action of one foot or the other. Intermittently, the bird compresses its feathers and pauses to peck dust against itself and under the body and, at least in the gallinaceous bird, to lower and rub its head and neck in the dust. On ceasing to dust, the bird stands, ruffles its plumage and vigorously shakes out all dust particles. Unlike bathing, dusting is characteristic of comparatively few birds. It appears to have developed commonly among birds of deserts and grasslands. Some birds–e.g., wrens house sparrows (*Passer domesticus*)–bathe as well as dust; others such as gallinaceous birds and goatsuckers dust only, rarely if ever taking a bath by deliberately standing in water. Dusting no doubt removes ectoparasites and may have other functions as yet undetermined.

**Anting**

A behaviour in which the fluids from ants, notably formic acid, are applied to the plumage and possibly the skin. There are two basic types of the behaviour : *active anting,* in which the bird anoints its plumage with the bill and *passive anting,* in which the bird allows ants to invade and anoint the plumage. *Whitaker* has reviewed the subject and described active anting in the Orchard Oriole (*Icterus spurius*). In a typical performance, the oriole crushed the ants and dabbed them, with rapidly vibrating head, to the ventral surface of the wing tips and especially the under tail coverts and bases of the rectrices. It anointed the feathers of the belly and tibiae only occasionally. A regular feature of the behaviour was tripping and tumbling, the result of bringing the tail forward beneath the body.

In passive anting, as described by Simmons, "birds squat or lie down among the ants, often while applying them directly and going through the motions of doing so with empty bill, while the wings and tail are partly or wholly spread out." Whitaker listed 148 species, 16 of them non-passerine, which have been captivity, perform active anting with burning matches and cigarette smoke, moth balls, hair tonic, mustard, citrus fruits, causes anting and what purposes or purposes it serves, if any, need investigation. *V.B. Dubinin* discovered that the formic acid from ants may destroy feather mites and thus he believes that anting functions as a defense against them.

**Comfort Movements**

Any numbers of actions which presumably give the bird "comfort" Some common movements are *feather settling* by first raising the feathers, shaking the body, flapping the wings, and then depressing the feathers in to proper position; *stretching*, often by extending the leg and wing of one side while fanning the tail on that side, then repeating the action on the other side, *yawning*, opening and closing the free-moving mandible, in most birds the lower one; and *resting*, by standing on one or both feet, or by squatting down, the plumage relaxed, the head with the bill tucked under the scapulars in a sleeping posture.

**Sleeping**

Birds ordinarily sleep with the head retracted and turned so that it rests on the back with the bill tucked under the scapulars. But there is both interspecific and interspecific variation. Some individuals may sleep with the head retracted it the shoulders on which it seems to rest and the bill forward. Several familial groups, such as pigeons, normally sleep this way. If not sleeping when floating on water, birds usually squat rather than stand. If squatting on a perch, the skeletal and muscular arrangement of the hindlimbs is such that the toes automatically grasp and hold to a twig or branch. Goat-suckers, though lacking this arrangement, can sleep in trees by squatting with the body lengthwise on a wide branch.

**Selecting Habitat and Shelter**

Most species of birds select distinct habitats. This allows an efficient use of natural environment and assists in reducing competition between species. Birds no doubt recognize their habitats by visible cues, but what are the cues and how do preferences become established in the first place? To answer the question, one may identify the components of an environment and then determine how the preference develops by comparing individual birds raised in captivity under different conditions with wild trapped birds. *Klopfer* did just that with Chipping Sparrows (*Spizella passerina*), taken from mixed stands of pine and oak in the North Carolina Piedmont and kept in a chamber where they were exposed to both fresh pine and oak foliage. Since *Klopfer* designed the study to measure the bird's preference of foliage for perch sites, all other environmental variables except foliage and light were minimized as much as possible.

*Klopfer* found that adults, trapped in the wild, and young, reared in isolation with no foliage, both preferred pine over oak; but the young, reared in isolation with oak leaves, preferred oak over pine. Their choices suggest that size and shape of leaves and to a less extent light-shadow patterns are the cues and that the preference of one foliage over the other, though innate, may be modified by experience. *Klopfer's* study does not suggest that the selection of a habitat constitutes an innate response only to such simple features as size and shape of foliage. Almost certainly the selection depends on many cues, single or in combination. However, Klopfer's work does quite properly imply that birds, while innately preferring a particular habitat, are nonetheless genetically equipped to re-adapt to others and thus demonstrate considerable plasticity or *opportunism.* Field observations support this. For example, in northern Michigan the wood Thrush (*Hylocichla mustelina*) prefers cedar-spruce bogs to the deep deciduous woods one would expect. In west Virginia the Magnolia Warbler (*Dendroica magnolia*) appears in oak forests where as northward in North America it is ordinarily partial to spruce and hem lock forests.

Klopfer's study as well as field observations tend to support the concept that birds breeding in the temperature areas of North America show opportunism because their environment is unstable and they must cope with variability. Birds breeding in tropical areas less opportunistic because their environment is comparatively stable. The habitats which birds select must meet requirements of shelter favourable to their survival, day and night, in all seasons of their occupancy. Besides shelter that provides escape and protection from predators, there must be shelter for roosting . Sometimes one shelter may serve two or more purpose–e.g., a shelter. But whether serving one or more than one purpose, birds are nevertheless as selective of these as they are of the overall habitat. A few remarks about roosting shelter are appropriate since most works on ornithology give relatively attention to the subject.

*Klimstra* and *Ziccardi* found that covies of Bobwhites *(Colinus virginianus)* in southern Illinois much preferred to roost on well-drained ground at medium to low elevations facing south or southwest and protected by a low, sparse, canopy of vegetation. Any shift in roosting ground was mostly from lower to higher elevations, except in windy and inclement weather when the shift was to poorly drained sites at lower elevation. Some cavity-nesting birds may use their own nests for roosting. *Berger* watched a male Downy Woodpecker (*Dendrocopos pubescens*) in October excavate a roosting cavity in a dead naple, much as it would a nesting cavity, and use it regularly through the winter until late April. *Frazier* and *Nolan* noted five to 14 Eastern Bluebirds (*Sialia Sialis*) roosting together in a bird box during the winter in Indiana. Many other observes in such enclosures or even natural cavities. After a heavy fall of snow the Ruffed Grouse (*Bonasa Umbellus*) will plunge into the snow and pass the night.

The above illustrations bring out several important points. Besides providing protections from predators, the roosting shelter gives protection from unfavourable weather and helps conserve body heat. When two or more birds choose to roost together in a small cavity, they naturally conserve heat. Also, in selecting roosting shelter, birds show opportunism as, for example, when choosing a nestbox instead of a natural cavity or when taking advantage of a snowfall.

## Selecting Food and Feeding

Most species of birds have the structure and motor coordinations for taking a wide variety of food although they usually learn to select particulars items. Among closely allied species in the same habitat this reduces competition and assures each species its share of total food supply. Each species is innately equipped with fixed action patterns for seeking food and preparing it for consumption. In the young bird these movements may be elicited initially by somewhat generalized stimuli. Gallinaceous chicks, foe example, begin seeking food by pecking at small objects with colour values contrasting with the background. To test the food pecking responses of young Indian Peafowl (*Pavo cristatus*) to objects of different sizes and colours against different background colours, *Dilger* and *Wallen* exposed four newly hatched chicks to circular objects ranging in size from two to 14 millimeters, painted in six different colours, and placed against backgrounds with the same six colours.

The chicks preferred the smallest objects when given a choice of all sizes, presented in different colours in random combination with differently colours background. They also preferred the combinations of objects and backgrounds that offered medium rather than maximum or minimum contrast that one might expect, because edible objects in the wild are more usually in medium contrast with the background. *Dilger* and *Wallen* suggest that maximum contrasts may be "somewhat inhibiting or possibly even somewhat frightening." The size of the bill and its special adaptations are among the important factors in determining the food items a bird learns to select.

In general, young birds select smaller items than older birds; as their bills harden and their motor coordinations mature, they take larger items though not beyond the point where they can handle them efficiently. Provided they have the necessary behavioural and structural equipment, which in many species included the feet, birds eventually learn from their own trial-and-error responses ti generalized stimuli, or from watching their parents and conspecific associates, to select particular items in various local situations. In England since 1911, titmice and other small birds–at least 11 species altogether–have learned the habitat of drinking from milk bottles by either puncturing metal foil caps, or stripping off or removing paper or cardboard caps. Some observes reported titmice in groups following milkcarts and attacking bottles immediately upon delivery. While many titmice no doubt learned the habit from other titmice, it is just as likely that it developed independently among others.

**Playing**

Birds perform playful activities though not to the extent that mammals do. Young birds still in the nest show *exploratory pecking*–gently pecking one another and the nesting materials Chicks in a brood, after leaving the nest, similarly peck one anther and peck at objects on the ground. Occasionally, they pick up morsels of food and run with them, eating them later. Gull chicks when scarcely half-grown pick up, run about with, and drop sticks. Plants stalks, and other objects. All young birds, well in advance of flying, fan their wings from time to time. Such activities constitute what amounts to "practicing" and may be a form of learning to forage, escape, or fly. Fledglings and chicks of land species commonly make short flights or runs, turning sharply–an activity called *play-fleeing.*

Aquatic chicks give a comparable performance on the water by running, splashing, and turning on the surface, even attempting to dive. Adults of some species perform activities that have all the appearances of play and seem to serve no function other than the release of pent-up energy. At the Falkland Islands in the South Atlantic, *Pettingill* observed groups of Magellanic Penguins (*Spheniscus magellanicus*) running from a beach into the sea, quickly submerging, torpedoing away from shore in a wide circle, and as quickly emerging and hustling up the beach. Parrots are notoriously playful. The Kea (*Nestor notabilis*), a large powerful mountain species in New Zealand, pulls windshield wipers off cars tosses empty tin cans about for no apparent reason than to be doing something.

## Social Behaviour

Social behaviour involves the many kinds of interactions or joint activities that have to do with two or more individuals in the same species or even individuals in different species.

### Agonistic Behaviour

One of the most common forms of social behaviour is *agonistic behaviour* which includes all manner of hostile activities or displays from over attack to overt escape. Perhaps most obvious hostile displays are those that have become ritualized as social releasers. The most abundant of these are probably the *threat displays,* "designed " to produce a "frightening" effect, forcing an opponent to retreat. Threat displays are thought to be activated by drives to attack and escape. For example, a bird approaching the territorial boundary of another exhibits attack, but, as it crosses the boundary, the motivation to escape sets in, coming into balance with the motivation to attack.

Meanwhile, the owner of the territory, instead of being intimidated, proceeds to attack the intruder. If, at this point, the attack drive of the intruder still continuous more between the two birds results; if the escape driver of the intruder becomes strong enough to inhibit or out balance its attack drive, the bird flees. Threat displays are believed to be derived from a number of different sources, among them the unritualized intention movements associated with locomotion. In certain passerine birds the threat displays may include lowering the body to the horizontal, facing and extending the head toward the opponent, sleeking the head feathers, holding the wings stiffly down or perhaps flitting them, depressing and fanning the tail or sometimes flipping it. All such actions suggest the movements of a bird about to take flight.

Almost as common as threat displays are the *appeasement displays* serving to prevent attack without provoking escape. These displays serve to reduce the strength of opponent's drive to attack and, to a lesser extent, the opponents drive to attack and, to lesser extent, the opponents drive to escape. As in threat displays, most of the movements have been derived from unritualized hostile sources, in this case, mostly from intention movements involving escape. These include, among certain passerine birds, sitting, flexing the legs, "hunching" with the head held low and retracted to the shoulders and turning away from the opponent.

Vocalization frequently complement both threat and appeasement displays, giving them an auditory component. Most songs of male birds function in part as threat displays, warning and/ or repelling rival males. Threat displays serve to space out birds, thereby preventing over-population in any one area. Appeasement displays, on the other hand, birds closer together by inhibiting attack or aggression and promoting bonds between pairs for reproduction and between parents and young.

## Defense Behaviour

All birds are subjects to predation and, consequently, show behavioural adaptations that promote their survival. Like maintenance behaviours, the defense behaviour of the individual is directed toward itself, in this case its own safety, but it may also serve to promote the survival of other individuals–e.g., the individual's own young, also the individuals of its own species and/or other species. Thus defense behaviour, under given circumstances, can be a form of social behaviour. Birds respond to certain danger stimuli both auditory and visual. Unaccustomed sharp sounds and abrupt movements evoke defensive behaviour of some sort, but if the same sounds are repeated again and again with further intensification, the defensive responses subside through habituation.

Models of owls that have the right combination of shape, colour pattern, contour can elicit alarm reactions in small birds. Cardboard models, cut in the shape of a short-necked, avian predator flying overhead and passed over the birds in the direction of the short neck, can induce escape reactions in waterfowl. The exact shape of the wings and tail and the colour are immaterial. Although it was once though the response to models were simply innate, more recent researches have demonstrated that at least some learning is involved.

The precise manner in which birds respond to danger signals depends on the physical and behavioural adaptations of the species responding and the type of predator, the intensity of signals, whether or not the responding individual is breeding, and the circumstances of the encounter. Some of the more common responses are the following:

## Fleeing and Freezing

If the average bird is exposed without benefit of cover and the danger signal, such as a hawk in flight, is strong, the bird flees to the nearest cover where it is likely to *"freeze"* –i.e., to keep completely motionless with the feathers sleeked, the legs, the head down in line with the body but cocked so that one eyes is fixed on the direction of the predator. If the bird is already in cover when receiving the danger signal, it freezes where it is. The freezing posture is concealing and, at the same time, practically identical to take off posture; thus bird is prepared to flee if necessary.

Birds such as grouse, snipe, woodcock, and goatsuckers with cryptic colouration respond to a strong danger signal by freezing even when not in a shelter. Most chicks, nestlings, and fledglings of all species normally freeze in response to a danger stimulus and often close their eyes. Non-breeding sandpipers, plovers, small gulls, terns, sifts, and other shelter-free dwellers, capable of fast or erratic flight, on the appearance of a flying predator, may attempt to outside or outmaneuver it should it attack.

## Communicating Danger

The vocal repertories of practically all birds include special alarm or warning calls given in response to danger signals, but not necessarily in all cases or at all times. A passerine bird at the sight of a cat invariably utters alarm calls, yet the same bird may freeze at the sight of a flying hawk and remain silent or given alarm calls only if it has young. Though presumably not consciously given and mates and induce freezing among young.

## Threatening

Confronted with potential predators or animal forms that impose danger birds may respond with actions that threaten or intimidate. In some instances the actions may appear identical to the threat displays against members of their own species, but often they are more exaggerated or more elaborate. Threatening behaviour includes actions that make the performer seem more formidable.

Thus most birds puff up their feathers, appearing larger than they are; and a great many birds open their mouths, as if threatening to bite, and lift their wings, as if intending to strike. Some birds utter infrequent sounds such as hisses that may function as an element of surprise and deter an approach.

A Great Horned Owl (*Bubo virginianus*) not only hisses but also snaps its bill rapidly; a Least Bittern (*Ixobrychus exilis*) silently lifts its bill skyward and sways back and forth. Birds with young most often employ threatening behaviour and may vary the type and intensity with kind of predator. A bird that performs vigorously at the approach of a small mammal may flee at the approach of a hawk or a human being.

**Attacking**

Many birds with nests or young, instead of threatening a predator, may attack it outright. If threatening behaviour fails ti intimidate a predator, attacking behaviour may follow. In a sense, attacking behaviour is a projection of threatening behaviour.

**Exploring**

Occasionally, birds show a tendency to explore strange objects such as blind suddenly placed near their nest. While examining it visually from different angels, they exhibit what appears to be a conflict between the urge to fly away (escape) and the urge to see more. Exploratory behaviour of this sort may have survival value by leading to the discovery of incipient danger.

**Mobbing**

In any season, groups of small birds respond to a perched hawk or owl by *"mobbing"* it. During the breeding season, birds often leave their territories temporarily to join in the performance. The basic motivation may be a combination of two tendencies : to explore and to escape. In some instances, the tendency to escape may be supplanted by the tendency to attack outright. The function of mobbing is puzzling. Possibly it serves to put all birds in a given area on the alert and, during the breeding season, to distract a predator from discovering young birds in the vicinity.

**Giving Distraction Displays**

Parent birds, threatened by a predator, may perform movements that sometimes stimulate incapacity from injury ("*injury-feigning*") or sometimes suggest the running away of a small mammal (“*rodent-running*”). The motivation is believed to be a combination of two crives–to escape and to attack–that was become ritualized and functions in diverting the attention of the predator from the nest or young to the parents. Injury-feigning shows up in a great number of species representing many families. The principal movements consist of fanning, beating, or dragging one or both wings, spreading and depressing the tail, fluffing the back and rump feathers, giving “distress’ calls, and alternately moving parallel with and away from the predator.

The performance differs in detail from species to species nesting on the ground most fully perform injury-feigning, but species nesting in other situations, such as tress and marshes, at least use some of the movements and adapt them accordingly. Rodent-running is characteristic os species nesting on the ground in thick grassy cover. Essentially it amounts to the bird sneaking off the nest, away from the predator, and running fast, occasionally hoping slightly and even fluttering its wings as it disappears from sight. Full distraction displays at maximum intensity are given when the eggs are hatching and for one or two days thereafter. While some predators may elicit distraction displays in a species, others may instead elicit outright. I a displaying bird is harried by a predator or a human being for an extensive period, its performance gradually wanes in intensity and may cease altogether.

## Flocking Behaviour

The majority of birds tend to be gregarious at one time or another. During the breeding season, most sea birds (including pelicans, cormorants, gulls and terns) and certain species of herons, swallows, and tropical icterids form *colonies* in which pairs establish and maintain nesting territories close to one another. Outside the breeding season a great many birds gather in *flocks* in which they remain until they disperse to nest. A few birds, such as the Monk Parakeet (*myiopsitta monachus*) of Argentina and certain species of weaver finches (Ploceidae) of Africa, are gregarious at all times, nesting in colonies and staying in flocks when not breeding. Flocks may be composed of one species or several species of similar size with approximately the same flight speeds but not necessarily the same food preferences and feeding adaptations. If a flock is of one species, as is more often the case, it may be comprised of one family or multiples of families from the preceding breeding season. This is believed true with small flocks of migrating geese.

*Fig. 8.2. Predator Mobbing. A noisy flock of crow mobs a golden eagle.*

Indeed, at Delta, Manitoba, colour-banded adult and young -of-the-year Canada Geese (*Branta canadensis*) were seen " to depart southward together in November and return, still as a united group, the next April." This same make-up may also be true of small flocks of permanent-resident chickadees (*Parus spp.*) during the winter. But there are numerous exceptiones. For examples, among shore birds migration southward from; birds-of-the -year come in later flocks. Early spring flocks of Red-winged. Blackbirds (*Agelaius phoeniceus*) are likely to be made up entirely of males. Each bird in a flock very rarely, expect when roosting for the night, comes in physical contact with its fellows, but maintains its *individual distance.* Cliff Swallows (*Petrochelidon pyrrhonota*), perched on a wire, keep about four inches apart. Any individual attempting to sit closer is immediately threatened by the birds it is crowding.

Numerous signals such as peculiar vocalizations and colour marks maintain the integrity of a flock. To these signals each individual responds. Should an individual become separated from the flock, it quickly shows appetitive behaviour by calling and searching until it finds its flock members. With in a closely integrated flock all the activities–feeding, preening, bathing, etc.– and movements tend to occur simultaneously. Usually one individual initiates the activities and the others soon perform them too–a procedure called *social facilitation* because the performance of one individual stimulates the behaviour of its fellows to act similarly. Through social facilitation a bird will sometimes eat more than its normal capacity by seeing one or more of its associates feeding. The concerted almost instantaneous movements in a flock are attributed to a "following" reaction, the instinctive urge better example of this trait in operation than in a compact group of wheeling over the area in perfect unison; and settling again, still compact.

In small flocks of conspecific birds, whether caged or in the wild, a social arrangement may be established that is normally hierarchial with each individual having its own rank in descending order of dominance. *Schjelderup-Ebbe* first brought this phenomenon to light in the Domestic Fowl (*Gallus gallus*). Studying penned birds, he discovered the existence of a peck order wherein Bird A can peck all the others and is therefore dominant, Bird B can peck all but Bird A, Bird C can peck all but Birds A and B, so on until there is one bird that can peck no other and is subordinate to all. Instead of a direct-line peck order, there can be a triangular or polygon arrangement in which Bird A pecks B, B pecks C, and C pecks A. This situation is apt to be only temporary and may revert back to the direct-line hierarchy. The position, rank, or *peck right* of the bird in the hierarchy is determined by one or a combination of factors such as sex, vigor, stage of molt, colour, and health. If something happens to the bird–for example, if it is injured and thus physically impaired–it may lose its peck right in the hierarchy and become the subordinates bird. Should the bird at the top position. A newcomer to a flock with only by starting as the subordinate bird.

In some flocks of conspecific birds the peck order is less fixed. Instead of a bird having a peck right over another and making that bird an absolute subordinate, it may achieve a *peck dominance* over another by winning more encounters than it loses. This may be observed in flocks of chickadees or Tree Sparrows (*Spizella arborea*) at a feeding station. Among heterosexual flocks of birds there may be more than one hierarchy. In a caged flock of 12 Red Crossbills (*Loxia curvirostra*), *Tordoff* noted three hierarchies, a peck order of males, a peck order of females, and a peck dominance of males over females. The lowest-ranking male was most active in

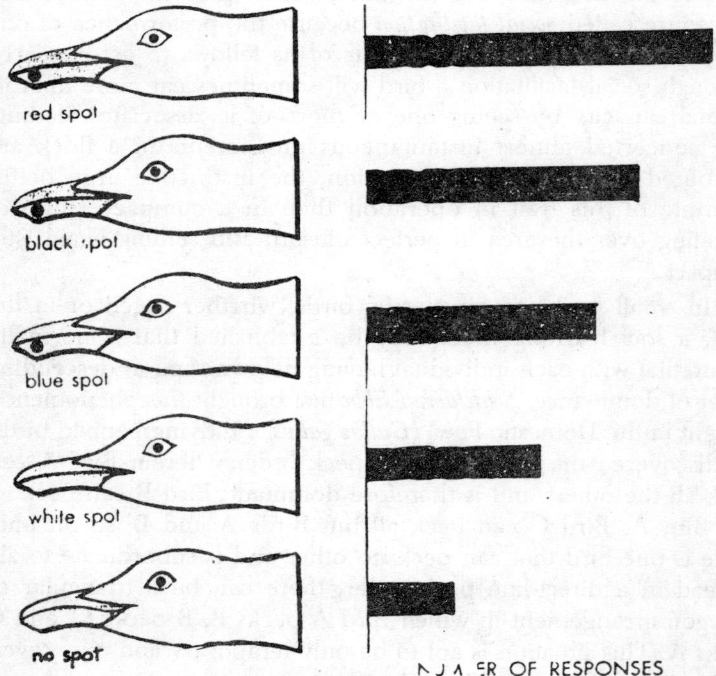


*Fig. 8.3. Herring-gull nestlings tap at the red spot on the bill of the parent; the parent then feeds them. N. Tinbergen presented cardboard models of the adult's head to newly hatched chicks and obtained the results indicated in the graph. Evidently, innate behaviour depends upon the nature of the stimulus.*

dominating females, followed closely by the top-ranking male. In a wild flock of Ring-necked Pheasants (*Phasianus colchicus*), *Collias* and *Taber* found a dominated all the hens until the most onset of the breeding season when they stopped pecking them.

Heterospecific flocks of birds show a species hierarchy, generally established in accordance with size and or hostility of the species comprising it. *Colquhoun,* for instance, observed that, when flocks of nuthatches and titmice were feeding, the European Nuthatch (*Sitta europaea*) usually dominated the Great Tit (*Parus major*), the Great Tit dominated the Blue Tit (*P. caeruleus*), and the Blue Tit dominated the Marsh Tit (*P. palustris*). Whether or not a flock has leadership remains a most question A family flock of Canada Geese is commonly believed to be led by the adult male, and a migrating wedge of the same species made up of multiple families is thought to be led by one of the older males, but all this is supposition.

*Hanson* found that the adult male Canada Goose will often initiate a local flights of his family but when the family is on the ground or in the water, the female often leads the flock while the male guards it at the rear. If there is a leadership in most flocks of adult birds, there is no evidence that it is taken by the top-ranking birds in the hierarchy. Quite possibly any persisting leadership of a flock of fully mature birds is non-existent since, as *Edmund Selous* once suggested, the flock is an individual, acting, moving, and behaving as a "collective bird." Flocking behaviour has survival value in at two important respects: (1) Birds together with integrated behaviour are less vulnerable to predators than solitary individuals. With more eyes to watch, the flocks has a greater awareness of the enemy's approach. When threatened, flock members acting on concert can take deterrent action. Either they can draw themselves together in a formidable mass as ducks resting on a pond converge to form a tight raft, or they can attack with a mobbing action as do gulls and terns in a nesting colony. (2) Birds foraging in a group increases the chances of success in finding and exploiting sources of food.

## Reproductive Behaviour

The interactions of birds that have to do with territory, mating, nesting, and parental care of the young come under the heading of reproductive behaviour, treated in later sections of this book.

## General Study Procedures

The science of behaviour, or *ethology,* is essentially the study of animal activities by direct observation, supplemented by experimental methods. While ethology is not exactly new, it does represent "return-to-nature" approach in understanding the function of the live animal. Although much is being learned rapidly about the behaviour of birds, was gaps in knowledge remain, even of the most common species. Avian ethology is still a wide open field for investigation. In studying the behaviour of any bird species, the student must describe and name its activities or behaviours. In other words, he must know thoroughly the behavioural make-up of his subject and be able to designate the different components.

### Describing Behaviours

Describing a behaviour is much more difficult than describing form, shape, colour, and morphological features. As *Dilger* stated, behaviour is ephemeral–it takes place and is gone. It may last a fraction of a second, or, like sleeping, it may last for hours. Moreover, a behaviour not only varies normally as does structure, but it also varies in the individual bird, depending on the subjects physiological state, motivation and many other factors. An adequate description of a behaviour must, therefore, be based on many more samples than would be required for structure. Commonly the difficulty in describing behaviour is the lack of suitable terms and consequent recourse to words that describe human activities. This frequently results, however unitentionally, in ascribing to a bird certain human traits and motives.

In any science treatment of behaviour always try to avoid anthropomorphisms–i.e., statements that suggest human thoughts, attitudes, and reactions–and concepts that involve teleology–the doctrine that gives a conscious purpose or objective to any undertaken. Also compounding the difficulty in describing a behaviour is its complexity, no matter how simple it may seem. Any behaviour may often be comprised of two or more separate *acts,* each of which is a set of observable activities, seen in components (*Russell* et al., 1954).

*Dilger* described a good example of a particular behaviour and its component acts in the Peach-faced Lovebird (*Agapornis roseicollis*) which carries stripes of nesting material in the feathers of its lower back and rump. When the bird puts a strip of paper in the feathers, first it *points* the stripe either to the right or to the left. Next, the bird *tucks* the strip into the feathers of the lower back by turning its head over one shoulder and bringing its bill with the strip in to the feathers, which are simultaneously *raised.* Then the bird *thrusts* the strip hard into the feathers with a *trembling* motion of its head, *releases* the strip, and *brings its head forward* while *compressing* the feathers. In all, Dilger noted at least eight separate acts performed in a characteristic manner and sequence. At the outset of his observations he considered thrusting and trembling a single act but, with later experiments, he found them to be separate because some birds trembled without thrusting.

## Naming Behaviours

The established custom is to derive names that serve to describe or at least indicate the most prominent action. If the behaviour is a display–i.e., a behaviour that has evolved as a signal or releaser–the name is usually capitalized as, for example, wings-up or bill-up-and-open, but if the behaviour is of any other sort, the name is not capitalized–e.g., bill down or tremble-thrust. As a rule, it is inadvisable to include in the name of a behaviour any word which designates its function or factors responsible for it–e.g., wings-up *threat* or *aggressive* bill-up-and-open. Such an interpretation is unwarranted, may eventually be proved wrong, and may bias unwarranted, amy eventually be proved wrong, and amy bias future investigations.

## Objective of the Study

Having described the behaviours of a bird, the student may proceed to postulate their function, survival value, causes, and origin.

## Function

What a behaviour accomplishes. It may simple serve the individual performer, as in the case of a maintenance activity or in a flight from one place to another, or it may, in the case of a display, serve to elicit a response in another bird. One can usually postulate a function of a display by nothing the kind of response elicited from the bird before which the display is performed.

## Survival Value

Any normal behaviour in an individual bird is presumed to enhance the welfare of its species, otherwise the behaviour would not have evolved. The survival value of biological significance of many behaviours–e.g., flying to a food source, escaping from a predator, etc.–may be so obvious it needs no attention, or it may be obscure. In that case, numerous observations of the behaviour, performed repeatedly under varying circumstances in the wild and/or in captivity, may provide the clue or at least the basis for making a reasonable guess. Why the occasional singing at night by certain species that regularly sing during the day? Why the teetering and tail-pumping action among many different species? Questions of this sort present a challenge to the avian ethologist.

**Causes**

A bird normally performs a behaviour from both internal and sexual cause. Internal causes include drive, usually conditioned by experience or learning; external causes may be one or more sign stimuli arising from the environment. The student can often determine external causes by careful field observations. In the natural environmental the situation in which a behaviour occurs is rarely the same from one occasion to the next. Thus it is possible for the student to establish the cause by observing and comparing a long series of occasions, in which the behaviour is not performed, and nothing the occasions, in which the behaviour is nor performed. In a sense, this a "natural environment." For example, what causes small birds to be alarmed by hawks on some occasions, the student will discover that small birds generally show alarm when the hawk is in flight, but not when it is perched. The movement of the bird in flight is the principle cause of the alarm. Another way of studying the external cause of behaviours is to compare similar behaviours apparently acquired independently by unrelated species–e.g., the habit developed by Cattle Egrets (*Bubulcus ibis*) and Brown-headed Cow birds (*Molothrus ater*) of following large, grazing mammals in search of insects stirred up by their feet.

There is always the chance that a comparative study of a habit common on very different birds will yield the clue to its common cause. Establishing the internal causes of behaviour and the role of experience in performing it requires sophisticated procedures under laboratory conditions, One method, to determine the role of experience, is the so-called deprivation experiment in which two groups of birds are used, one experimental, the other a control. The groups are identical expect that one group has been deprived of experience. By comparing subsequent changes in the performance of the experimental group to the "base line" of behaviour of the control group, one can determine the extent to which experience plays a part in the behaviour.

**Origin**

How behaviours have evolved is an important objective in ethological studies. After describing and observing all the behaviours in a species, the student may postulate that certain displays are derived from maintenance activities, others from displacement activities, intention movements, and so on. By comparing certain homologous behaviours in closely related bird species, the student may discern different stages in the evolution of these behaviours-information that may also shed light on the evolution of the species themselves.